CURRENT INTELLIGENCE BULLETIN 62

Asbestos Fibers and Other Elongate Mineral Particles: State of the Science and Roadmap for Research

DEPARTMENT OF HEALTH AND HUMAN SERVICES
Centers for Disease Control and Prevention
National Institute for Occupational Safety and Health

Cover Photograph: Transitional particle from upstate New York identified by the United States Geological Survey (USGS) as anthophyllite asbestos altering to talc. Photograph courtesy of USGS.

> This document is in the public domain and may be freely copied or reprinted.

Disclaimer

Mention of any company or product does not constitute endorsement by the National Institute for Occupational Safety and Health (NIOSH). In addition, citations to Web sites external to NIOSH do not constitute NIOSH endorsement of the sponsoring organizations or their programs or products. Furthermore, NIOSH is not responsible for the content of these Web sites.

Ordering Information

To receive NIOSH documents or other information about occupational safety and health topics, contact NIOSH at

   Telephone: 1–800–CDC–INFO (1–800–232–4636)
   TTY: 1–888–232–6348
   E-mail: cdcinfo@cdc.gov

or visit the NIOSH Web site at www.cdc.gov/niosh.

For a monthly update on news at NIOSH, subscribe to *NIOSH eNews* by visiting www.cdc.gov/niosh/eNews.

DHHS (NIOSH) Publication No. 2011–159

(Revised for clarification; no changes in substance or new science presented)

April 2011

SAFER • HEALTHIER • PEOPLE™

Foreword

Asbestos has been a highly visible issue in public health for over three decades. During the mid- to late-20th century, many advances were made in the scientific understanding of worker health effects from exposure to asbestos fibers and other elongate mineral particles (EMPs). It is now well documented that asbestos fibers, when inhaled, can cause serious diseases in exposed workers. However, many questions and areas of confusion and scientific uncertainty remain.

The National Institute for Occupational Safety and Health (NIOSH) has determined that exposure to asbestos fibers causes cancer and asbestosis in humans on the basis of evidence of respiratory disease observed in workers exposed to asbestos, and recommends that exposures be reduced to the lowest feasible concentration. As the federal agency responsible for conducting research and making recommendations for the prevention of worker injury and illness, NIOSH has undertaken a reappraisal of how to ensure optimal protection of workers from exposure to asbestos fibers and other EMPs. As a first step in this effort, NIOSH convened an internal work group to develop a framework for future scientific research and policy development. The NIOSH Mineral Fibers Work Group prepared a first draft of this *State of the Science and Roadmap for Scientific Research* (herein referred to as the *Roadmap*), summarizing NIOSH's understanding of occupational exposure and toxicity issues concerning asbestos fibers and other EMPs.

NIOSH invited comments on the occupational health issues identified and the framework for research suggested in the first draft *Roadmap*. NIOSH sought other views about additional key issues that should be identified, additional research that should be conducted, and methods for conducting the research. In particular, NIOSH sought input from stakeholders concerning study designs, techniques for generating size-selected fibers, analytic approaches, sources of particular types of EMPs suitable for experimental studies, and worker populations suitable for epidemiological study. On the basis of comments received during the public and expert peer review process, NIOSH revised the *Roadmap* and invited public review of the revised version by stakeholders. After further revision and public comment, a revised draft *Roadmap* was submitted for review by the National Academies of Science in early 2009. Based on the National Academies assessment of the draft *Roadmap*, revisions were made, and NIOSH disseminated a fourth draft version of the document for final public comment in early 2010. After considering these comments, NIOSH has developed this final revision of the *Roadmap*.

The purpose of this *Roadmap* is to outline a research agenda that will guide the development of specific research programs and projects that will lead to a broader and clearer understanding of the important determinants of toxicity for asbestos fibers and other EMPs. NIOSH recognizes that results from such research may impact environmental as well as occupational health policies and practices. Many of the issues that are important in the workplace are also important to communities and to the general population. Therefore, NIOSH envisions that the planning and conduct of the research will be a collaborative effort involving active participation of multiple federal agencies, including the Agency for Toxic Substances and Disease Registry (ATSDR), the Consumer Product Safety Commission (CPSC), the Environmental Protection Agency (EPA), the Mine Safety and Health Administration (MSHA), the National Institute of Environmental Health Sciences (NIEHS), the National Institute of Standards and Technology (NIST), the National Toxicology Program (NTP), the Occupational Safety and Health Administration (OSHA), and the United States Geological Survey (USGS), as well as labor, industry, academia, health and safety practitioners, and other interested parties, including international groups. This collaboration will help to focus the scope of the research, to fund and conduct the research, and to develop and disseminate informational materials describing research results and their implications for establishing new occupational and public health policies.

This *Roadmap* also includes a clarified rewording of the NIOSH recommended exposure limit (REL) for airborne asbestos fibers. This clarification is not intended to establish a new NIOSH occupational health policy for asbestos, and no regulatory response by OSHA or MSHA is requested or expected.

John Howard, MD
Director, National Institute for
  Occupational Safety and Health
Centers for Disease Control and Prevention

Executive Summary

In the 1970s, NIOSH determined that exposure to asbestos fibers causes cancer and asbestosis in humans on the basis of evidence of respiratory disease observed in workers exposed to asbestos. Consequently, it made recommendations to the federal enforcement agencies on how to reduce workplace exposures. The enforcement agencies developed occupational regulatory definitions and standards for exposure to airborne asbestos fibers based on these recommendations. Since the promulgation of these standards, which apply to occupational exposures to the six commercially used asbestos minerals—the serpentine mineral chrysotile, and the amphibole minerals cummingtonite-grunerite asbestos (amosite), riebeckite asbestos (crocidolite), actinolite asbestos, anthophyllite asbestos, and tremolite asbestos—the use of asbestos in the United States has declined substantially and mining of asbestos in the United States ceased in 2002. Nevertheless, many asbestos products remain in use and new asbestos-containing products continue to be manufactured in or imported into the United States.

As more information became available on the relationship between the dimensions of asbestos fibers and their ability to cause nonmalignant respiratory disease and cancer, interest increased in exposure to other "mineral fibers." The term "mineral fiber" has been frequently used by nonmineralogists to encompass thoracic-size elongate mineral particles (EMPs) occurring either in an asbestiform habit (e.g., asbestos fibers) or in a nonasbestiform habit (e.g., as needle-like [acicular] or prismatic crystals), as well as EMPs that result from the crushing or fracturing of nonfibrous minerals (e.g., cleavage fragments). Asbestos fibers are clearly of substantial health concern. Further research is needed to better understand health risks associated with exposure to other thoracic-size EMPs, including those with mineralogical compositions identical or similar to the asbestos minerals and those that have already been documented to cause asbestos-like disease, as well as the physicochemical characteristics that determine their toxicity.

Imprecise terminology and mineralogical complexity have affected progress in research. "Asbestos" and "asbestiform" are two commonly used terms that lack mineralogical precision. "Asbestos" is a term used for certain minerals that have crystallized in a particular macroscopic habit with certain commercially useful properties. These properties are less obvious on microscopic scales, and so a different definition of asbestos may be necessary at the scale of the light microscope or electron microscope, involving characteristics such as chemical composition and crystallography. "Asbestiform" is a term applied to minerals with a macroscopic habit similar to that of asbestos. The lack of precision in these terms and the difficulty in translating macroscopic properties to microscopically identifiable characteristics contribute to miscommunication and uncertainty in identifying toxicity associated with various forms of minerals. Deposits may have more than one mineral habit and transitional minerals may be present, which make it difficult to clearly and simply describe the mineralogy.

In 1990, NIOSH revised its recommendation concerning occupational exposure to airborne asbestos fibers. At issue were concerns about potential health risks associated with worker exposures to the analogs of the asbestos minerals that occur in a different habit—so-called cleavage fragments—and the inability of the analytical method routinely used for characterizing airborne exposures (i.e., phase contrast microscopy [PCM]) to differentiate nonasbestiform analogs from asbestos fibers on the basis of physical appearance. This problem was further compounded by the lack of more sensitive analytical methods that could distinguish asbestos fibers from other EMPs having the same elemental composition. To address these concerns and ensure that workers are protected, NIOSH defined "airborne asbestos fibers" to encompass not only fibers from the six previously listed asbestos minerals (chrysotile, crocidolite, amosite, anthophyllite asbestos, tremolite asbestos, and actinolite asbestos) but also EMPs from their nonasbestiform analogs. NIOSH retained the use of PCM for measuring airborne fiber concentrations and counting those EMPs having: (1) an aspect ratio of 3:1 or greater and (2) a length greater than 5 μm. NIOSH also retained its recommended exposure limit (REL) of 0.1 airborne asbestos fiber per cubic centimeter (f/cm^3).

Since 1990, several persistent concerns have been raised about the revised NIOSH recommendation. These concerns include the following:

- NIOSH's explicit inclusion of EMPs from nonasbestiform amphiboles in its 1990 revised definition of airborne asbestos fibers is based on inconclusive science and contrasts with the regulatory approach subsequently taken by OSHA and by MSHA.

- The revised definition of airborne asbestos fibers does not explicitly encompass EMPs from asbestiform amphiboles that formerly had been mineralogically defined as tremolite (e.g., winchite and richterite) or other asbestiform minerals that are known to be (e.g., erionite and fluoro-edenite) or may be (e.g., some forms of talc) associated with health effects similar to those caused by asbestos.

- The specified dimensional criteria (length and aspect ratio) for EMPs covered by the revised definition of airborne asbestos fibers may not be optimal for protecting the health of exposed workers because they are not based solely on health concerns.

- Other physicochemical parameters, such as durability and surface activity, may be important toxicological parameters but are not reflected in the revised definition of airborne asbestos fibers.

- NIOSH's use of the term "airborne asbestos fibers" to describe all airborne EMPs covered by the REL differs from the way mineralogists use the term and this inconsistency leads to confusion about the toxicity of EMPs.

NIOSH recognizes that its 1990 description of the particles covered by the REL for airborne asbestos fibers has created confusion, causing many to infer that the nonasbestiform minerals included in the NIOSH definition are "asbestos." Therefore, in this *Roadmap*, NIOSH makes clear that such nonasbestiform minerals are not "asbestos" or "asbestos minerals," and clarifies which particles are included in the REL. This clarification also provides a basis for a better understanding of the need for the proposed research. Clarification of the REL in this way does not change the existing NIOSH occupational health policy for asbestos, and no regulatory response by OSHA or MSHA is requested or expected. The REL remains subject to revision based on findings of ongoing and future research.

PCM, the primary method specified by NIOSH, OSHA, and MSHA for analysis of air samples for asbestos fibers, has several limitations, including limited ability to resolve very thin fibers and to differentiate various types of EMPs. Occupational exposure limits for asbestos derive from lung cancer risk estimates from exposure of workers to airborne asbestos fibers in commercial processes. These risk assessments are based on fiber concentrations determined from a combination of PCM-based fiber counts on membrane filter samples and fiber counts estimated from impinger samples. The standard PCM method counts only fibers longer than 5 µm. Moreover, some fibers longer than 5 µm may be too thin to be detected by PCM. Thus, an undetermined number of fibers collected on each sample remain uncounted by PCM. More sensitive analytical methods are currently available, but standardization and validation of these methods will be required before they can be recommended for routine analysis. However, unlike PCM, these methods are substantially more expensive, and field instruments are not available. In addition, any substantive change in analytical techniques used to evaluate exposures to asbestos and/or the criteria for determining exposure concentrations will necessitate a reassessment of current risk estimates, which are based on PCM-derived fiber concentrations.

Epidemiological evidence clearly indicates a causal relationship between exposure to fibers from the asbestos minerals and various adverse health outcomes, including asbestosis, lung cancer, and mesothelioma. However, NIOSH has viewed as inconclusive the results from epidemiological studies of workers exposed to EMPs from the nonasbestiform analogs of the asbestos minerals. Populations of interest for possible epidemiological studies include workers at talc mines in upstate New York and workers at taconite mines in northeastern Minnesota. Others include populations exposed to other EMPs, such as winchite and richterite fibers (asbestiform EMPs identified in vermiculite from a former mine near Libby, Montana), zeolites (such as asbestiform erionite), and other minerals (such as fluoro-edenite). Future studies should include detailed characterizations of the particles to which workers are or have been exposed.

There is considerable potential for experimental animal and *in vitro* studies to address specific scientific questions relating to the toxicity of EMPs. Short-term *in vivo* animal studies and *in vitro* studies have been conducted to examine cellular and tissue responses to EMPs, identify pathogenic mechanisms involved in those responses, and understand morphological and/or physicochemical EMP properties controlling those mechanisms. Long-term studies of animals exposed to EMPs have been conducted to assess the risk for adverse health outcomes (primarily lung cancer, mesothelioma, and lung fibrosis) associated with various types and dimensions of EMPs. Such studies have produced evidence demonstrating the importance of dimensional characteristics of mineral particles for determining carcinogenic potential of durable EMPs. Although *in vitro* studies and animal studies are subject to uncertainties with respect to how their findings apply to humans, such studies are warranted to systematically assess and better understand the impacts of dimension, morphology, chemistry, and biopersistence of EMPs on malignant and nonmalignant respiratory disease outcomes.

To reduce existing scientific uncertainties and to help resolve current policy controversies, a strategic research program is needed that encompasses endeavors in toxicology, exposure assessment, epidemiology, mineralogy, and analytical methods. The findings of such research can contribute to the development of new policies for exposures to airborne asbestos fibers and other EMPs with recommendations for exposure indices that are not only more effective in protecting workers' health but firmly based on quantitative estimates of health risk. To bridge existing scientific uncertainties, this *Roadmap* proposes that interdisciplinary research address the following three strategic goals: (1) develop a broader and clearer understanding of the important determinants of toxicity for EMPs, (2) develop information on occupational exposures to various EMPs and health risks associated with such exposures, and (3) develop improved sampling and analytical methods for asbestos fibers and other EMPs.

Developing a broader and clearer understanding of the important determinants of toxicity for EMPs will involve building on what is known by systematically conducting *in vitro* studies and *in vivo* animal studies to ascertain which physical and chemical properties of EMPs influence their toxicity and their underlying mechanisms of action in causing disease. The *in vitro* studies could provide information on membranolytic, cytotoxic, and genotoxic activities as well as signaling mechanisms. The *in vivo* animal studies should involve a multispecies testing approach for short-term assays to develop information for designing chronic inhalation studies and to develop information on biomarkers and mechanisms of disease. Chronic animal inhalation studies are required to address the impacts of dimension, morphology, chemistry, and biopersistence on critical disease endpoints, including cancer and nonmalignant respiratory disease. Chronic inhalation studies should be designed to provide solid scientific evidence on which to base human risk assessments for a variety of EMPs. The results of toxicity studies should be assessed in the context of results of epidemiologic studies to provide a basis for understanding the human health effects of exposure to EMPs for which epidemiologic studies are not available.

Developing information and knowledge on occupational exposures to various EMPs and potential health outcomes will involve (1) collecting and analyzing available occupational exposure information to ascertain the characteristics and extent of exposure to various types of EMPs; (2) collecting and analyzing available information on health outcomes associated with exposures to various types of EMPs; (3) conducting epidemiological studies of workers exposed to various types of EMPs to better define the association between exposure and health effects; and (4) developing and validating methods for screening, diagnosis, and secondary prevention for diseases caused by exposure to asbestos fibers and other EMPs.

Developing improved sampling and analytical methods for EMPs will involve (1) reducing interoperator and interlaboratory variability of currently used analytical methods; (2) developing a practical analytical method that will permit the counting, sizing, and identification of EMPs; (3) developing a practical analytical method that can assess the potential durability of EMPs as one determinant of biopersistence in the lung; and (4) developing and validating size-selective sampling methods for collecting and quantifying airborne thoracic-size asbestos fibers and other EMPs.

A primary anticipated outcome of the research that is broadly outlined in this *Roadmap* will be the identification of the physicochemical parameters such as chemical composition, dimensional attributes (e.g., ranges of length, width, and aspect ratio), and durability as predictors of biopersistence, as well as particle surface characteristics or activities (e.g., generation reactive oxygen species [ROS]) as determinants of toxicity of asbestos fibers and other EMPs. The results of the research will also help define the sampling and analytical methods that closely measure the important toxic characteristics and need to be developed. These results can then inform development of appropriate recommendations for worker protection.

Another outcome of the research that is broadly outlined in this *Roadmap* might be the development of criteria that could be used to predict, on the basis of results of *in vitro* testing and/or short-term *in vivo* testing, the potential risk associated with exposure to any particular type of EMP. This could reduce the need for comprehensive toxicity testing with long-term *in vivo* animal studies and/or epidemiological evaluation of each type of EMP. Ultimately, the results from such studies could be used to fill in knowledge gaps beyond EMPs to encompass predictions of relative toxicities and adverse health outcomes associated with exposure to other elongate particles (EPs), including inorganic and organic manufactured particles. A coherent risk management approach that fully incorporates an understanding of the toxicity of particles could then be developed to minimize the potential for disease in exposed individuals and populations.

This *Roadmap* is intended to outline the scientific and technical research issues that need to be addressed to ensure that workers are optimally protected from health risks posed by exposures to asbestos fibers and other EMPs. Achievement of the research goals framed in this *Roadmap* will require a significant investment of time, scientific talent, and resources by NIOSH and others. This investment, however, can result in a sound scientific basis for better occupational health protection policies for asbestos fibers and other EMPs.

Contents

Foreword iii
Executive Summary v
List of Figures xiv
List of Tables xiv
Abbreviations xv
Acknowledgments xvii
1 Introduction 1
2 Overview of Current Issues 5
   2.1 Background 5
   2.2 Minerals and Mineral Morphology 6
   2.3 Terminology 7
        2.3.1 Mineralogical Definitions 8
        2.3.2 Other Terms and Definitions 9
   2.4 Trends in Asbestos Use, Occupational Exposures, and Disease 9
        2.4.1 Trends in Asbestos Use 9
        2.4.2 Trends in Occupational Exposure 10
        2.4.3 Trends in Asbestos-related Disease 12
   2.5 Workers' Home Contamination 14
   2.6 Clinical Issues 15
   2.7 The NIOSH Recommendation for Occupational Exposure
        to Asbestos 17
        2.7.1 The NIOSH REL as Revised in 1990 18
        2.7.2 Clarification of the Current NIOSH REL 33
   2.8 EMPs Other than Cleavage Fragments 34
        2.8.1 Chrysotile 34
        2.8.2 Asbestiform Amphibole Minerals 36
        2.8.3 Other Minerals of Potential Concern 38
   2.9 Determinants of Particle Toxicity and Health Effects 39
        2.9.1 Deposition 39
        2.9.2 Clearance and Retention 40

		2.9.3	Biopersistence and other Potentially Important Particle Characteristics...............................	42
		2.9.4	Animal and *In Vitro* Toxicity Studies	46
		2.9.5	Thresholds	58
	2.10	Analytical Methods..		60
		2.10.1	NIOSH Sampling and Analytical Methods for Standardized Industrial Hygiene Surveys.............	62
		2.10.2	Analytical Methods for Research	63
		2.10.3	Differential Counting and Other Proposed Analytical Approaches for Differentiating EMPs	65
	2.11	Summary of Key Issues...		66
3	Framework for Research			69
	3.1	Strategic Research Goals and Objectives........................		70
	3.2	An Approach to Conducting Interdisciplinary Research......		70
	3.3	National Reference Repository of Minerals and Information System		70
	3.4	Develop a Broader Understanding of the Important Determinants of Toxicity for Asbestos Fibers and Other EMPs.................		71
		3.4.1	Conduct *In Vitro* Studies to Ascertain the Physical and Chemical Properties that Influence the Toxicity of Asbestos Fibers and Other EMPs	76
		3.4.2	Conduct Animal Studies to Ascertain the Physical and Chemical Properties that Influence the Toxicity of Asbestos Fibers and Other EMPs	78
		3.4.3	Evaluate Toxicological Mechanisms to Develop Early Biomarkers of Human Health Effects...............	81
	3.5	Develop Information and Knowledge on Occupational Exposures to Asbestos Fibers and Other EMPs and Related Health Outcomes		81
		3.5.1	Assess Available Information on Occupational Exposures to Asbestos Fibers and Other EMPs	82
		3.5.2	Collect and Analyze Available Information on Health Outcomes Associated with Exposures to Asbestos Fibers and Other EMPs	83
		3.5.3	Conduct Selective Epidemiological Studies of Workers Exposed to Asbestos Fibers and Other EMPs.............	84
		3.5.4	Improve Clinical Tools and Practices for Screening, Diagnosis, Treatment, and Secondary Prevention of Diseases Caused by Asbestos Fibers and Other EMPs......	87

3.6 Develop Improved Sampling and Analytical Methods for
Asbestos Fibers and Other EMPs 88

    3.6.1 Reduce Inter-operator and Inter-laboratory Variability of
the Current Analytical Methods Used for Asbestos Fibers .. 90

    3.6.2 Develop Analytical Methods with Improved Sensitivity to
Visualize Thinner EMPs to Ensure a More Complete
Evaluation of Airborne Exposures 91

    3.6.3 Develop a Practical Analytical Method for Air Samples
to Differentiate Asbestiform Fibers from the Asbestos
Minerals and EMPs from Their Nonasbestiform Analogs .. 92

    3.6.4 Develop Analytical Methods to Assess Durability
of EMPs 93

    3.6.5 Develop and Validate Size-selective Sampling Methods
for EMPs 93

3.7 From Research to Improved Public Health Policies for
Asbestos Fibers and Other EMPs 94

4 The Path Forward 99

4.1 Organization of the Research Program 99
4.2 Research Priorities 100
4.3 Outcomes 101

5 References 103

6 Glossary 131

6.1 NIOSH Definition of Potential Occupational Carcinogen 131
6.2 Definitions of New Terms Used in this Roadmap 131
6.3 Definitions of Inhalational Terms 131
6.4 Definitions of General Mineralogical Terms and Specific Minerals ... 132
6.5 References for Definitions of General Mineralogical Terms,
Specific Mineral and Inhalation Terms 132

List of Figures

Figure 1. U.S. asbestos production and imports, 1991–2007

Figure 2. Asbestos: Annual geometric mean exposure concentrations by major industry division, MSHA and OSHA samples, 1979–2003

Figure 3. Number of asbestosis deaths, U.S. residents aged ≥ 15 years, 1968–2004

Figure 4. Number of deaths due to malignant mesothelioma, U.S. residents aged ≥ 15 years, 1999–2005

List of Tables

Table 1. Definitions of general mineralogical terms.

Table 2. Definitions of specific minerals.

Abbreviations

8-OHdG	8-hydroxydeoxyguanosine
AED	aerodynamic equivalent diameter
AIHA	American Industrial Hygiene Association
AP-1	activator protein-1
ASTM	ASTM International [previously American Society for Testing and Materials]
ATSDR	Agency for Toxic Substances Disease Registry
BAL	bronchoalveolar lavage
BrdU	bromodeoxyuridine
CI	confidence interval
COX-2	cyclooxygenase-2
CPSC	Consumer Product Safety Commission
DM	dark-medium microscopy
DNA	deoxyribonucleic acid
DPPC	dipalmitoyl phosphatidylcholine
ED	electron diffraction
EDS	energy dispersive X-ray spectroscopy
EGFR	epidermal growth factor receptor
EM	electron microscopy
EMP	elongate mineral particle
EP	elongate particle
EPA	U.S. Environmental Protection Agency
ERK	extracellular signal-regulated kinase
ESR	electron spin resonance
f/cm^3	fibers per cubic centimeter
f/mL-yr	fibers per milliliter-year
HSL/ULO	Health and Safety Laboratory/UL Optics
ICD	International Classification of Diseases
IgG	immunoglobulin G
IL	interleukin
IMA	International Mineralogical Association
IMIS	Integrated Management Information System
IP	intraperitoneal
ISO	International Organization for Standardization
L	liter
LDH	lactate dehydrogenase
LOQ	limit of quantification
MDH	Minnesota Department of Health
mg/m^3-d	milligrams per cubic meter-days

MAPK	mitogen-activated protein kinase
MMAD	mass median aerodynamic diameter
MMMF	man-made mineral fiber
MMVF	man-made vitreous fiber
mppcf	million particles per cubic foot
MSHA	Mine Safety and Health Administration
mRNA	messenger ribonucleic acid
NADPH	nicotinamide adenine dinucleotide phosphate
NFκB	nuclear factor kappa beta
NIEHS	National Institute of Environmental Health Sciences
NMRD	nonmalignant respiratory disease
NIOSH	National Institute for Occupational Safety and Health
NIST	National Institute of Standards and Technology
NORA	National Occupational Research Agenda
NORMS	National Occupational Respiratory Mortality System
NTP	National Toxicology Program
OSHA	Occupational Safety and Health Administration
PCMe	phase contrast microscopy equivalent
PCM	phase contrast microscopy
PEL	permissible exposure limit
RCF	refractory ceramic fiber
REL	recommended exposure limit
ROS	reactive oxygen species
RTV	RT Vanderbilt Company, Inc.
SAED	selected area X-ray diffraction
SEM	scanning electron microscopy
SMR	standardized mortality ratio
SO	superoxide anion
SOD	superoxide dismutase
SV40	simian virus 40
SVF	synthetic vitreous fiber
SWCNT	single-walled carbon nanotubes
TEM	transmission electron microscopy
TF	tissue factor
TGF	transforming growth factor
TNF-α	tumor necrosis factor-alpha
TWA	time-weighted average
USGS	United States Geological Survey
WHO	World Health Organization
XPS	X-ray photoelectron spectroscopy

Acknowledgments

This document was prepared under the aegis of the NIOSH Mineral Fibers Work Group by members of the NIOSH staff. Many internal NIOSH reviewers not listed also provided critical feedback important to the preparation of this *Roadmap*.

The NIOSH Mineral Fibers Work Group acknowledges the contributions of Jimmy Stephens, PhD, former NIOSH Associate Director for Science, who initiated work on this document and articulated many of its most critical issues in an early draft.

The NIOSH Mineral Fibers Work Group also acknowledges the contributions of Gregory Meeker, USGS, who participated in discussions of the pertinent mineralogy and mineralogical nomenclature.

NIOSH Mineral Fibers Work Group

Paul Baron, PhD
John Breslin, PhD
Robert Castellan, MD, MPH
Vincent Castranova, PhD
Joseph Fernback, BS
Frank Hearl, SMChE
Martin Harper, PhD

Jeffrey Kohler, PhD
Paul Middendorf, PhD, Chair
Teresa Schnorr, PhD
Paul Schulte, PhD
Patricia Sullivan, ScD
David Weissman, MD
Ralph Zumwalde, MS

Major Contributors

Paul Middendorf, PhD
Ralph Zumwalde, MS
Robert Castellan, MD, MPH
Martin Harper, PhD
William Wallace, PhD

Leslie Stayner, PhD
Vincent Castranova, PhD
Frank Hearl, SMChE
Patricia Sullivan, ScD

Peer Reviewers

NIOSH greatly appreciates the time and efforts of expert peer reviewers who provided comments and suggestions on the initial publicly disseminated draft of the *Roadmap* (February 7, 2007 version).

William Eschenbacher, MD
Group Health Associates

Morton Lippmann, PhD
New York University

David Michaels, PhD, MPH
George Washington University

Franklin Mirer, PhD
Hunter College

Brooke Mossman, PhD
University of Vermont

Brad Van Gosen, MS
U.S. Geological Survey

L. Christine Oliver, MD, MPH
Harvard School of Medicine

Ann Wylie, PhD
University of Maryland

William N. Rom, MD, MPH
New York University

Institutes of Medicine and National Research Council of the National Academy of Sciences

NIOSH appreciates the time and efforts of the National Academy of Sciences (NAS) committee members, consultant, and study staff who contributed to the development of the NAS report *A Review of the NIOSH Roadmap for Research on Asbestos Fibers and Other Elongate Mineral Particles* on the January 2009 version of the draft *Roadmap*. The individuals contributing to the report are identified in the NAS document.

Document History

Throughout its development, this *Roadmap* has undergone substantial public comment and scientific peer review with subsequent revision. All external input has been considered and addressed, as appropriate, to ultimately produce this final version of the *Roadmap*. A listing of the various draft versions disseminated for public comment and/or scientific peer review is presented here.

February 2007—Draft entitled *Asbestos and Other Mineral Fibers: A Roadmap for Scientific Research* was disseminated for public comment and scientific peer review.

June 2008—Draft entitled *Revised Draft NIOSH Current Intelligence Bulletin—Asbestos Fibers and Other Elongate Mineral Particles: State of the Science and Roadmap for Research* was disseminated for public comment.

January 2009—Draft entitled *Revised Draft NIOSH Current Intelligence Bulletin—Asbestos Fibers and Other Elongated Mineral Particles: State of the Science and Roadmap for Research* was submitted to the Institute of Medicine and the National Research Council of the National Academies of Science for scientific review.

January 2010—Draft entitled *Draft NIOSH Current Intelligence Bulletin—Asbestos Fibers and Other Elongate Mineral Particles: State of the Science and Roadmap for Research—Version 4* was disseminated for public comment.

March 2011—Final Version of *NIOSH Current Intelligence Bulletin—Asbestos Fibers and Other Elongate Mineral Particles: State of the Science and Roadmap for Research* was published.

1 Introduction

Many workers are exposed to a broad spectrum of inhalable particles in their places of work. These particles vary in origin, size, shape, chemistry, and surface properties. Considerable research over many years has been undertaken to understand the potential health effects of these particles and the particle characteristics that are most important in conferring their toxicity. Elongate particles* (EPs) have been the subject of much research, and the major focus of research on EPs has related to asbestos fibers, a group of elongate mineral particles (EMPs) that have long been known to cause serious disease when inhaled. Because of the demonstrated health effects of asbestos, research attention has also been extended not only to other EMPs, but also to synthetic vitreous fibers which have dimensions similar to asbestos fibers and, more recently, to engineered carbon nanotubes and carbon nanofibers. Although nonmineral EPs are of interest, they are not the subject of this *Roadmap*, which focuses on EMPs.

Occupational health policies and associated federal regulations controlling occupational exposure to airborne asbestos fibers have been in existence for decades. Current regulations have been based on studies of the health effects of exposures encountered in the commercial exploitation and use of asbestos fibers. Nevertheless, important uncertainties remain to be resolved to fully inform possible revision of existing federal policies and/or development of new federal policies to protect workers from health effects caused by occupational exposure to airborne asbestos fibers.

Health effects caused by other exposures to EMPs have not been studied as thoroughly as health effects caused by exposures in the commercial exploitation and use of asbestos fibers. Miners and others exposed to asbestiform amphibole fibers associated with vermiculite from a mine near Libby, Montana, may not have been exposed to commercial asbestos fibers, but the adverse health outcomes they have experienced as a result of their exposure have indicated that those EMPs are similarly toxic. Other hardrock miner populations face uncertain but potential risk associated with exposures to EMPs that could be generated during mining and processing of nonasbestiform amphiboles. Studies of human populations exposed to airborne fibers of erionite, a fibrous mineral that is neither asbestos nor amphibole, have documented high rates of malignant mesothelioma (a cancer most commonly associated with exposure to asbestos fibers). Further research is warranted to understand how properties of EMPs determine toxicity so that the nature and magnitude of any potential toxicity associated with an EMP to which workers are exposed can be readily predicted and

---

*A glossary of technical terms is provided in Section 6. It includes definitions of terms from several standard sources in Tables 1 and 2. In general, where a term is used in this *Roadmap* the definitions in Tables 1 and 2 under the column "NIOSH 1990" are intended unless otherwise specified. Readers should be aware that, in reviewing publications cited in this *Roadmap*, it was not always possible to know the meaning of technical terms as used by the authors of those publications. Thus, some imprecision of terminology carries over into the literature review contained in this *Roadmap*.

controlled, even when exhaustive long-term studies of that particular EMP have not been carried out.

This *Roadmap* has been prepared and is being disseminated with the intent of motivating eventual development and implementation of a coordinated, interdisciplinary research program that can effectively address key remaining issues relating to health hazards associated with exposure to asbestos fibers and other EMPs.

Section 2, *Overview of Current Issues*, provides an overview of available scientific information and identifies important issues that need to be resolved before recommendations for occupational exposure to airborne asbestos fibers and related EMPs can be improved and before recommendations for occupational exposure to other EMPs can be developed. The nature of occupational exposures to asbestos has changed over the last several decades. Once dominated by chronic exposures in asbestos textile mills, friction product manufacturing, cement pipe fabrication, and insulation manufacture and installation, occupational exposures to asbestos in the United States now primarily occur during maintenance activities or remediation of buildings containing asbestos. OSHA has estimated that 1.3 million workers in general industry continue to be exposed to asbestos. In 2002 NIOSH estimated that about 44,000 mine workers might be exposed to asbestos fibers or amphibole cleavage fragments during the mining of some mineral commodities. These current occupational exposure scenarios frequently involve short-term, intermittent exposures, and proportionately fewer long fibers than workers were exposed to in the past. The generally lower current exposures and predominance of short fibers give added significance to the question of whether or not there are thresholds for EMP exposure level and EMP length below which workers would incur no demonstrable risk of material health impairment. The large number of potentially exposed workers and these changed exposure scenarios also give rise to the need to better understand whether appropriate protection is provided by the current occupational exposure recommendations and regulations. In addition, limited information is currently available on exposures to, and health effects of, other EMPs.

Section 3, *Framework for Research*, provides a general framework for research needed to address the key issues. NIOSH envisions that this general framework will serve as a basis for a future interdisciplinary research program carried out by a variety of organizations to (1) elucidate exposures to EMPs, (2) identify any adverse health effects caused by these exposures, and (3) determine the influence of size, shape, and other physical and chemical characteristics of EMPs on human health. Findings from this research would provide a basis for determining which EMPs should be included in recommendations to protect workers from hazardous occupational exposures along with appropriate exposure limits. A fully informed strategy for prioritizing research on EMPs will be based on a systematic collection and evaluation of available information on occupational exposures to EMPs.

Section 4, *The Path Forward*, broadly outlines a proposed structure for development and oversight of a comprehensive, interdisciplinary research program. Key to this approach will be (1) the active involvement of stakeholders representing parties with differing views, (2) expert study groups specifying and guiding various components of the research program, and (3) a multidisciplinary group providing careful ongoing review and oversight to ensure relevance, coordination, and impact of the overall research program. NIOSH does not intend this (or any other) section of this *Roadmap* to be prescriptive, so detailed research aims, specific research priorities, and funding considerations have intentionally not been specified. Rather, it is expected that these more detailed aspects of the program will be most effectively developed with collaborative input from scientists, policy experts, and managers from various agencies, as well as from other interested stakeholders.

2 Overview of Current Issues

2.1 Background

Prior to the 1970s, concern about the health effects of occupational exposure to airborne fibers was focused on six commercially exploited minerals termed "asbestos:" the serpentine mineral chrysotile and the amphibole minerals cummingtonite-grunerite asbestos (amosite), riebeckite asbestos (crocidolite), actinolite asbestos, anthophyllite asbestos, and tremolite asbestos. The realization that dimensional characteristics of asbestos fibers were important physical parameters in the initiation of respiratory disease led to studies of other elongate mineral particles (EMPs) of similar dimensions [Stanton et al. 1981].

To date, occupational health interest in EMPs other than asbestos fibers has been focused primarily on fibrous minerals exploited commercially (e.g., wollastonite, sepiolite, and attapulgite) and mineral commodities that contain (e.g., Libby vermiculite) or may contain (e.g., upstate New York talc) asbestiform minerals. Exposure to airborne thoracic-size EMPs generated from the crushing and fracturing of nonasbestiform amphibole minerals has also garnered substantial interest. The asbestos minerals, as well as other types of fibrous minerals, are typically associated with other minerals in geologic formations at various locations in the United States [Van Gosen 2007]. The biological significance of occupational exposure to airborne particles remains unknown for some of these minerals and can be difficult to ascertain, given the mixed and sporadic nature of exposure in many work environments and the general lack of well-characterized exposure information.

The complex and evolving terminology used to name and describe the various minerals from which airborne EMPs are generated has led to much confusion and uncertainty in scientific and lay discourse related to asbestos fibers and other EMPs. To help reduce such confusion and uncertainty about the content of this *Roadmap*, several new terms are used in this *Roadmap* and defined in the Glossary (Section 6). The Glossary also lists definitions from a variety of sources for many other mineralogical and other scientific terms used in this *Roadmap*. Definitions from these sources often vary, and many demonstrate a lack of standardization and sometimes rigor that should be addressed by the scientific community.

To address current controversies and uncertainties concerning exposure assessment and health effects relating to asbestos fibers and other EMPs, strategic research endeavors are needed in toxicology, exposure assessment, epidemiology, mineralogy, and analytical methods. The results of such research might inform new risk assessments and the potential development of new policies for asbestos fibers and other EMPs, with recommendations for exposure limits that are firmly based on well-established risk estimates and that effectively protect workers' health. To support the development of these policies, efforts are needed to establish a common set of mineral terms that unambiguously describe EMPs and are relevant for toxicological assessments. What follows in the remainder of Section 2 is an overview of (1) definitions and terms relevant to asbestos fibers and other EMPs; (2) trends in production and use of asbestos; (3) occupational exposures to asbestos and asbestos-related diseases; (4) sampling and analytical issues; and (5) physicochemical properties associated with EMP toxicity.

2.2 Minerals and Mineral Morphology

Minerals are naturally occurring inorganic compounds with a specific crystalline structure and elemental composition. Asbestos is a term applied to several silicate minerals from the serpentine and amphibole groups that occur in a particular type of fibrous habit and have properties that have made them commercially valuable. The fibers of all varieties of asbestos are long, thin, and usually flexible when separated. One variety of asbestos, chrysotile, is a mineral in the serpentine group of sheet silicates. Five varieties of asbestos are minerals in the amphibole group of double-chain silicates—riebeckite asbestos (crocidolite), cummingtonite-grunerite asbestos (amosite), anthophyllite asbestos, tremolite asbestos, and actinolite asbestos [Virta 2002].

Although a large amount of health information has been generated on workers occupationally exposed to asbestos, limited mineral characterization information and the use of nonmineralogical names for asbestos have resulted in uncertainty and confusion about the specific nature of exposures described in many published studies. Trade names for mined asbestos minerals predated the development of rigorous scientific nomenclature. For example, amosite is the trade name for asbestiform cummingtonite-grunerite and crocidolite is the trade name for asbestiform riebeckite. A changing mineralogical nomenclature for amphiboles has also contributed to uncertainty in the specific identification of minerals reported in the literature. Over the past 50 years, several systems for naming amphibole minerals have been used. The current mineralogical nomenclature was unified by the International Mineralogical Association (IMA) under a single system in 1978 [Leake 1978] and later modified in 1997 [Leake et al. 1997]. For some amphibole minerals, the name assigned under the 1997 IMA system is different than the name used prior to 1978, but the mineral names specified in regulations have not been updated to correlate with the new IMA system names.

Adding to the complexity of the nomenclature, serpentine and amphibole minerals typically develop through the alteration of other minerals. Consequently, they may exist as partially altered minerals having variations in elemental compositions. For example, the microscopic analysis of an elongate amphibole particle using energy dispersive X-ray spectroscopy (EDS) can reveal variations in elemental composition along the particle's length, making it difficult to identify the particle as a single specific amphibole mineral. In addition, a mineral may occur in different growth forms, or "habits," so different particles may have different morphologies. However, because they share the same range of elemental composition and chemical structure and belong to the same crystal system, they are the same mineral. Different habits are not recognized as having different mineral names.

Mineral habit results from the environmental conditions present during a mineral's formation. The mineralogical terms applied to habits are generally descriptive (e.g., fibrous, asbestiform, massive, prismatic, acicular, asbestiform, tabular, and platy). Both asbestiform (a specific fibrous type) and nonasbestiform versions (i.e., analogs) of the same mineral can occur in juxtaposition or matrixed together, so that both analogs of the same mineral can occur within a narrow geological formation.

The habits of amphibole minerals vary, from prismatic crystals of hornblende through prismatic or acicular crystals of riebeckite, actinolite, tremolite, and others, to asbestiform habits of grunerite (amosite), anthophyllite, tremolite-actinolite, and riebeckite (crocidolite). The prismatic and acicular crystal habits occur more commonly, and asbestiform habit is relatively rare. Some of the amphiboles, such as hornblendes, are not known to occur in an asbestiform habit. The asbestiform varieties range from finer (flexible) to coarser (more brittle) and often are found in a mixture of fine and coarse fibrils. In addition, properties vary—e.g., density of (010) defects—even within an apparently homogeneous specimen [Dorling and Zussman 1987].

In the scientific literature, the term "mineral fibers" has often been used to refer not only to particles that occur in an asbestiform habit but also to particles that occur in other fibrous habits or as needle-like (acicular) single crystals. The term "mineral fibers" has sometimes also encompassed other prismatic crystals and cleavage fragments that meet specified dimensional criteria. Cleavage fragments are generated by crushing and fracturing minerals, including the nonasbestiform analogs of the asbestos minerals. Although the substantial hazards of inhalational exposure to airborne asbestos fibers have been well documented, there is ongoing debate about whether exposure to thoracic-size EMPs (EMPs of a size that can enter the thoracic airways when inhaled) from nonasbestiform analogs of the asbestos minerals is also hazardous.

2.3 Terminology

The use of nonstandard terminology or terms with imprecise definitions when reporting studies makes it difficult to fully understand the implications of these studies or to compare the results to those of other studies. For the health community, this ultimately hampers research efforts, leads to ambiguity in exposure-response relationships, and could also lead to imprecise recommendations to protect human health. Terms are often interpreted differently between disciplines. The situation is complicated by further different usage of the same terms by stakeholders outside of the scientific community. NIOSH has carefully reviewed numerous resources and has not found any current reference for standard terminology and definitions in several disciplines that is complete and unambiguous. An earlier tabulation of asbestos-related terminology by the USGS demonstrated similar issues [Lowers and Meeker 2002]. The terms "asbestos" and "asbestiform" exemplify this issue. They are commonly used terms but lack mineralogical precision. "Asbestos" is a term used for certain minerals that have crystallized in a particular macroscopic habit with certain commercially useful properties. These properties are less obvious on microscopic scales, and so a different definition of "asbestos" may be necessary at the scale of the light microscope or electron microscope, involving characteristics such as chemical composition and crystallography. "Asbestiform" is a term applied to minerals with a macroscopic habit similar to that of asbestos. The lack of precision in these terms and the difficulty in translating macroscopic properties to microscopically identifiable characteristics contribute to miscommunication and uncertainty in identifying toxicity associated with various forms of minerals. Some deposits may contain more than one habit or transitional particles may be present, which make it difficult to clearly and simply describe the mineralogy. Furthermore, the minerals included in the term asbestos vary between federal agencies and sometimes within an agency.

NIOSH supports the development of standard terminology and definitions relevant to the issues of asbestos and other EMPs that are based on objective criteria and are acceptable to the majority of scientists. NIOSH also supports the dissemination of standard terminology and definitions to the community of interested nonscientists and encourages adoption and use by this community. The need for the development and standardization of unambiguous terminology and definitions warrants a priority effort of the greater scientific community that should precede, or at least be concurrent with, further research efforts.

2.3.1 Mineralogical Definitions

The minerals of primary concern are the asbestiform minerals that have been regulated as asbestos (chrysotile, amosite,† crocidolite,‡ tremolite asbestos, actinolite asbestos, and anthophyllite asbestos). In addition, there is interest in closely related minerals to which workers might be exposed that (1) were not commercially used but would have been mineralogically identified as regulated asbestos minerals at the time the asbestos regulations were promulgated (e.g., asbestiform winchite and richterite); (2) are other asbestiform amphiboles (e.g., fluoro-edenite); (3) might resemble asbestos (e.g., fibrous antigorite); (4) are unrelated EMPs (e.g., the zeolites erionite and mordenite, fibrous talc, and the clay minerals sepiolite and palygorskite); and (5) are individual particles or fragments of the nonasbestiform analogs of asbestos minerals. Minerals are precisely defined by their chemical composition and crystallography. Ionic substitutions occur in minerals, especially for metal cations of similar ionic charge or size. Such substitution can result in an *isomorphous series* (also referred to as *solid-solution* or *mixed crystal*) consisting of minerals of varying composition between end-members with a specific chemical composition. The differences in chemical composition within an *isomorphous series* can result in different properties such as color and hardness, as well as differences in crystal properties by alteration of unit-cell dimensions. It is sometimes possible to differentiate mineral species on the basis of distinctive changes through an *isomorphous series*. However, in general, classification occurs by an arbitrary division based on chemistry, and this can be complicated by having multiple sites of possible substitution (e.g., in a specific mineral, calcium may exchange for magnesium in one position whereas sodium and potassium may be exchanged in another position). These allocations are open to reevaluation and reclassification over time (e.g., the mineral now named richterite was called soda-tremolite in pre-1978 IMA nomenclature).

When certain minerals were marketed or regulated as asbestos, the mineral names had definitions that might have been imprecise at the time and might have changed over time. In particular, the mineral name amosite was a commercial term for a mineral that was not well defined at first. The definitions of amosite in the *Dictionary of Mining, Mineral, and Related Terms* [USBOM 1996] and in the *Glossary of Geology* [American Geological Institute 2005] allow for the possibility that amosite might be anthophyllite asbestos, although it is now known to be a mineral in the cummingtonite-grunerite series. This is one source of confusion in the literature.

---

†Amosite is not recognized as a proper mineral name.
‡Crocidolite is not recognized as a proper mineral name.

A further source of confusion comes from the use of the geological terms for a mineral habit. Minerals of the same chemistry differing only in the expression of their crystallinity (e.g., massive, fibrous, asbestiform, or prismatic) are not differentiated in geology as independent species. Thus, tremolite in an asbestiform crystal habit is not given a separate name (either chemical or common) from tremolite in a massive habit. It has been suggested that crystals grown in an "asbestiform" habit can be distinguished by certain characteristics, such as parallel or radiating growth of very thin and elongate crystals that are to some degree flexible, the presence of bundles of fibrils, and, for amphiboles, a particular combination of twinning, stacking faults, and defects [Chisholm 1973]. The geological conditions necessary for the formation of asbestiform crystals are not as common as those that produce other crystal habits. These other habits may occur without any accompanying asbestiform crystals. However, amphibole asbestos may also include additional amphiboles that, if separated, are not asbestiform [Brown and Gunter 2003]. The mineralogical community uses many terms, including fibril, fiber, fibrous, acicular, needlelike, prismatic, and columnar, to denote crystals that are elongate. In contrast, in sedimentology, similar terms have been more narrowly defined with specific axial ratios.

Thus it is not clear, even from a single reference source, exactly what range of morphologies are described by these terms and the degree of overlap, if any. For example, the Dictionary of Mining, Mineral, and Related Terms defines fibril as "a single fiber, which cannot be separated into smaller components without losing its fibrous properties or appearance," but also defines a fiber as "the smallest single strand of asbestos or other fibrous material."

2.3.2 Other Terms and Definitions

Health-related professions also employ terminology that can be used imprecisely. For example, the terms "inhalable" and "respirable" have different meanings but are sometimes used interchangeably. Also, each of these terms is defined somewhat differently by various professional organizations and agencies. Particles can enter the human airways, but the aspiration efficiency, the degree of penetration to different parts of the airways, and the extent of deposition depend on particle aerodynamics, as well as on the geometry and flow dynamics within the airways. In addition to obvious differences between species (e.g., mouse, rat, dog, primate, human), there is a significant range of variation within a species based on, for example, age, sex, body mass, and work-rate. Thus, these terms may mean different things to a toxicologist engaging in animal inhalation experiments, an environmental specialist concerned with childhood exposure, and an industrial hygienist concerned with adult, mostly male, workers.

2.4 Trends in Asbestos Use, Occupational Exposures, and Disease

2.4.1 Trends in Asbestos Use

Over recent decades, mining and use of asbestos have declined in the United States. The mining of asbestos in the United States ceased in 2002. Consumption of raw asbestos continues to decline from a peak of 803,000 metric tons in 1973 [USGS 2006]. In 2006, 2000 metric tons of raw asbestos were imported, down from an estimated 35,000 metric tons in 1991 (see Figure 1) and a peak of 718,000 metric tons in 1973. Unlike information on the importation of raw asbestos, information is not

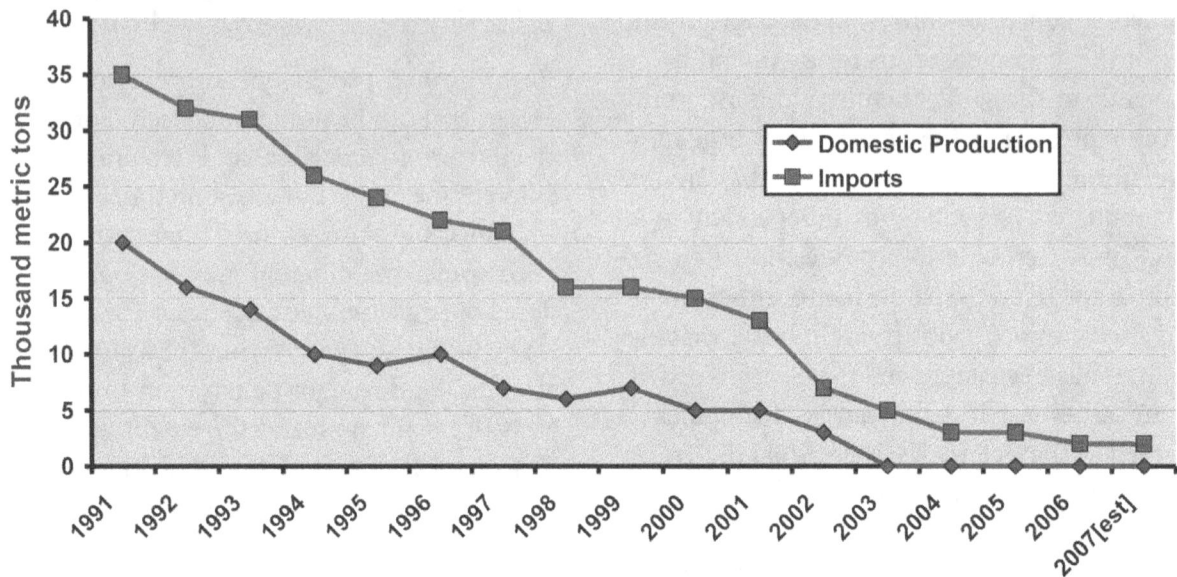

Figure 1. U.S. asbestos production and imports, 1991–2007. Source of data: USGS [2008].

readily available on the importation of asbestos-containing products. The primary recent uses for asbestos materials in the United States are estimated as 55% for roofing products, 26% for coatings and compounds, and 19% for other applications [USGS 2007], and more recently as 84% for roofing products and 16% for other applications [USGS 2008].

Worldwide, the use of asbestos has declined. Using the amount of asbestos mined as a surrogate for the amount used, worldwide annual use has declined from about 5 million metric tons in 1975 to about 2 million metric tons since 1999 [Taylor et al. 2006]. The European Union has banned imports and the use of asbestos with very limited exceptions. In other regions of the world, there is a continued demand for inexpensive, durable construction materials. Consequently, markets remain strong in some countries for asbestos-cement products, such as asbestos-cement panels for construction of buildings and asbestos-cement pipe for water-supply lines. Currently over 70% of all mined asbestos is used in Eastern Europe and Asia [Tossavainen 2005].

Historically, chrysotile accounted for more than 90% of the world's mined asbestos; it presently accounts for over 99% [Ross and Virta 2001; USGS 2008]. Mining of crocidolite (asbestiform riebeckite) and amosite (asbestiform cummingtonite-grunerite) deposits have accounted for most of the remaining asbestos. Mining of amosite is thought to have ceased in 1992 and mining of crocidolite is thought to have ended in 1997, although it is not possible to be certain. Small amounts of anthophyllite asbestos have been mined in Finland [Ross and Virta 2001] and are currently being mined in India [Ansari et al. 2007].

2.4.2 Trends in Occupational Exposure

Since 1986, the annual geometric mean concentrations of occupational exposures to asbestos in the United States, as reported in the Integrated Management Information System (IMIS) of the Occupational Safety and Health Administration

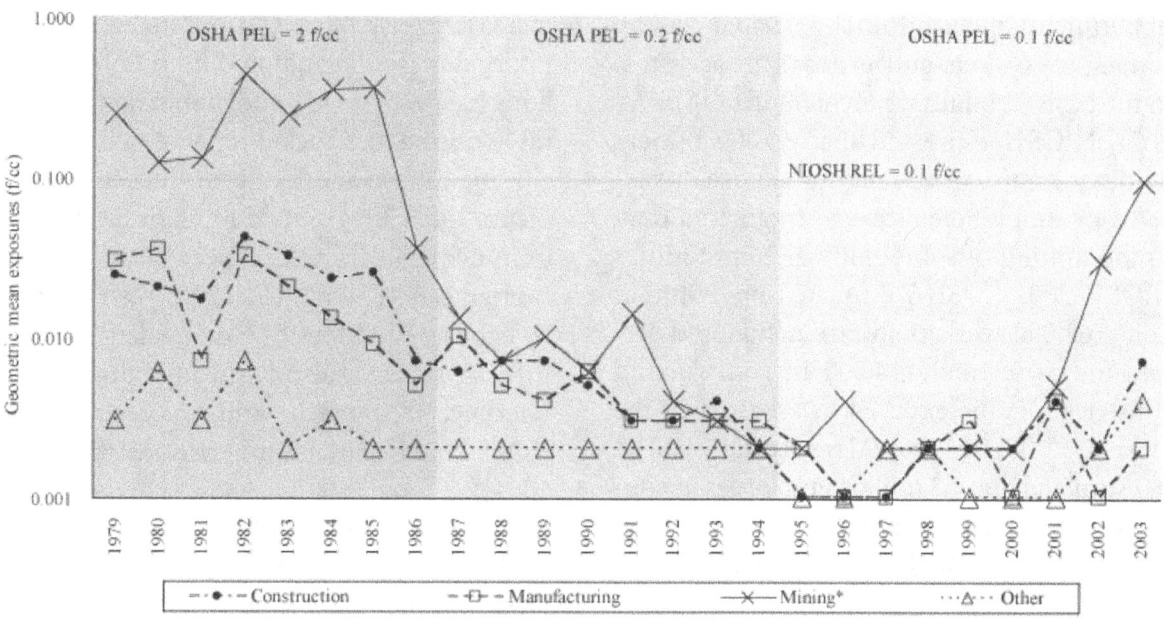

Figure 2. Asbestos: Annual geometric mean exposure concentrations by major industry division, MSHA and OSHA samples, 1979–2003. Source of data: NIOSH [2007a]. Note: the MSHA PEL for this time period was 2 f/cm³.

(OSHA) and the database of the Mine Safety and Health Administration (MSHA), have been consistently below the NIOSH recommended exposure limit (REL) of 0.1 fibers per cubic centimeter of air (f/cm³) for all major industry divisions (Figure 2). The number of occupational asbestos exposures that were measured and reported in IMIS decreased from an average of 890 per year during the 8-year period of 1987–1994 to 241 per year during the 5-year period of 1995–1999 and 135 for the 4-year period of 2000–2003. The percentage exceeding the NIOSH REL decreased from 6.3% in 1987–1994 to 0.9% in 1995–1999, but it increased to 4.3% in 2000–2003. During the same three periods, the number of exposures measured and reported in MSHA's database decreased from an average of 47 per year during 1987–1994 to an average of 23 per year during 1995–1999, but it increased to 84 during 2000–2003 (most of which were collected in 2000). The percentage exceeding the NIOSH REL decreased from 11.1% in 1987–1994 to 2.6% in 1995–1999, but it increased to 9.8% in 2000–2003 [NIOSH 2007a].

The preceding summary of occupational exposures to asbestos is based on the OSHA and MSHA regulatory definitions relating to asbestos. Because NIOSH includes nonasbestiform analogs of the asbestos minerals in the REL that may have, at least from some samples, been excluded by OSHA and MSHA in performing differential counting, the reported percentages of exposures exceeding the REL should be interpreted as lower limits. Because of analytical limitations of the phase contrast microscopy (PCM) method and the variety of workplaces from which the data were obtained, it is unclear what portions of these exposures were to EMPs from nonasbestiform analogs of the asbestos minerals, which have been explicitly encompassed by the NIOSH REL for airborne asbestos fibers since 1990.

Very limited information is available on the number of workers still exposed to asbestos. On the basis of mine employment data [MSHA 2002], NIOSH estimated that 44,000 miners and other mine workers may be exposed to asbestos or amphibole cleavage fragments during the mining of some mineral commodities [NIOSH 2002]. OSHA estimated in 1990 that about 568,000 workers in production and services industries and 114,000 in construction industries may be exposed to asbestos in the workplace [OSHA 1990]. More recently, OSHA has estimated that 1.3 million employees in construction and general industry face significant asbestos exposure on the job [OSHA 2008].

In addition to evidence from OSHA and MSHA that indicates a reduction in occupational exposures in the United States over the last several decades of the 1900s, other information compiled on workplace exposures to asbestos indicates that the nature of occupational exposures to asbestos has changed [Rice and Heineman 2003]. Once dominated by chronic exposures in manufacturing processes such as those used in textile mills, friction product manufacturing, and cement pipe fabrication, current occupational exposures to asbestos in the United States primarily occur during maintenance activities or remediation of buildings containing asbestos. These current occupational exposure scenarios frequently involve short-term, intermittent exposures.

2.4.3 Trends in Asbestos-related Disease

Evidence that asbestos causes lung cancer and mesothelioma in humans is well documented [NIOSH 1976; IARC 1977, 1987a,b; EPA 1986; ATSDR 2001; HHS 2005a]. Epidemiological studies of workers occupationally exposed to asbestos have clearly documented a substantial increase in risk of several nonmalignant respiratory diseases, including diffuse fibrosis of the lung (i.e., asbestosis) and nonmalignant pleural abnormalities including acute pleuritis and chronic diffuse and localized thickening of the pleura [ATS 2004]. In addition, it has been determined that laryngeal cancer [IOM 2006] and ovarian cancer [Straif et al. 2009] can be caused by exposure to asbestos, and evidence suggests that asbestos may also cause other diseases (e.g., pharyngeal, stomach, and colorectal cancers [IOM 2006] and immune disorders [ATSDR 2001]).

National surveillance data, showing trends over time, are available for two diseases with rather specific mineral fiber etiologies—asbestosis and malignant mesothelioma (see following subsections). Lung cancer is known to be caused in part by asbestos fiber exposure but has multiple etiologies. Ongoing national surveillance for lung cancer caused by asbestos exposure has not been done. However, using various assumptions and methods, several researchers have projected the number of U.S. lung cancer deaths caused by asbestos. Examples of the projected number of asbestos-caused lung cancer deaths in the United States include 55,100 [Walker et al. 1983] and 76,700 [Lilienfeld et al. 1988]; each of these projections represent the 30-year period from 1980 through 2009. However, in the absence of specific diagnostic criteria and a specific disease code for the subset of lung cancers caused by asbestos, ongoing surveillance cannot be done for lung cancer caused by asbestos.

2.4.3.1 Asbestosis

NIOSH has annually tracked U.S. deaths due to asbestosis since 1968 and deaths due to malignant mesothelioma since 1999, using death certificate data in the National Occupational Respiratory Mortality System (NORMS). NORMS

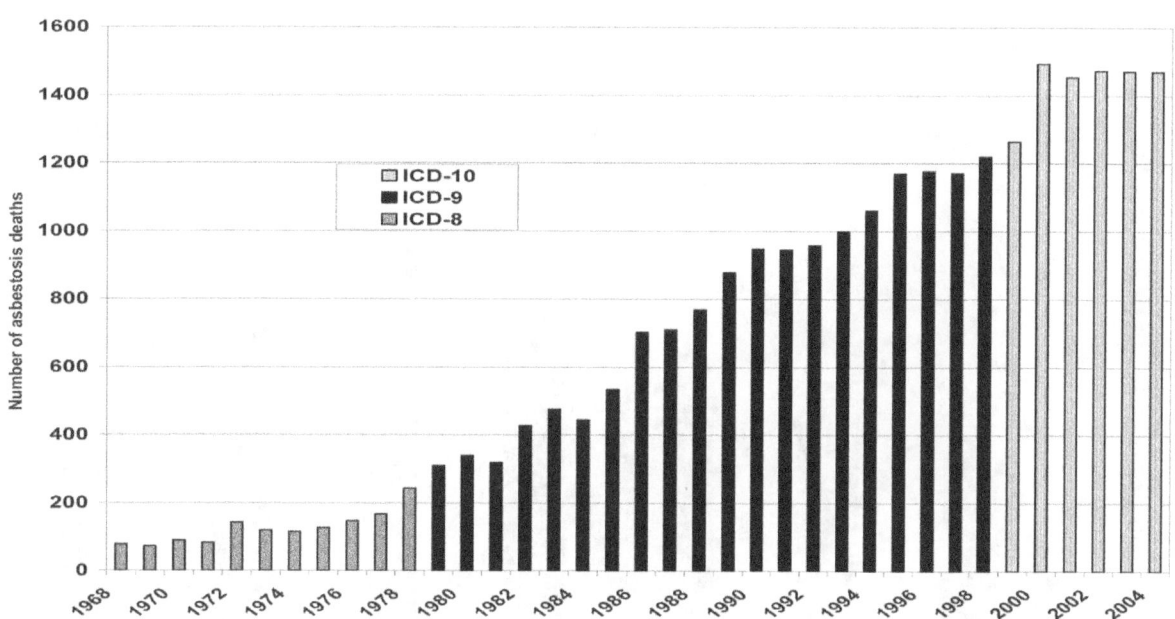

Figure 3. Number of asbestosis deaths, U.S. residents aged ≥15 years, 1968–2004. Source of data: NIOSH [2007b].

data, representing all deaths among U.S. residents, show that asbestosis deaths increased almost 20-fold from the late 1960s to the late 1990s (Figure 3) [NIOSH 2007b]. Asbestosis mortality trends are expected to substantially trail trends in asbestos exposures (see Section 2.4.2) for two primary reasons: (1) the latency period between asbestos exposure and asbestosis onset is typically long, commonly one or two decades or more; and (2) asbestosis is a chronic disease, so affected individuals can live for many years with the disease before succumbing. In fact, asbestosis deaths have apparently plateaued (at nearly 1,500 per year) since 2000 (Figure 3) [NIOSH 2007b]. Ultimately, it is anticipated that the annual number of asbestosis deaths in the United States will decrease substantially as a result of documented reductions in exposure. However, asbestos use has not been completely eliminated, and because asbestos-containing materials remain in structural materials and machinery, the potential for exposure continues. Thus, asbestosis deaths in the United States are anticipated to continue for several decades.

2.4.3.2 Malignant Mesothelioma

Malignant mesothelioma, an aggressive disease that is nearly always fatal, is known to be caused by exposure to asbestos and some other mineral fibers [IOM 2006]. The occurrence of mesothelioma has been strongly linked with occupational exposures to asbestos [Bang et al. 2006]. There had been no discrete International Classification of Disease (ICD) code for mesothelioma until its most recent 10th revision. Thus, only seven years of NORMS data are available with a specific ICD code for mesothelioma (Figure 4); during this period, there was a 9% increase in annual mesothelioma deaths, from 2,484 in 1999 to 2,704 in 2005 [NIOSH 2007b]. A later peak for mesothelioma deaths than for asbestosis deaths would be entirely expected, given the longer latency for mesothelioma

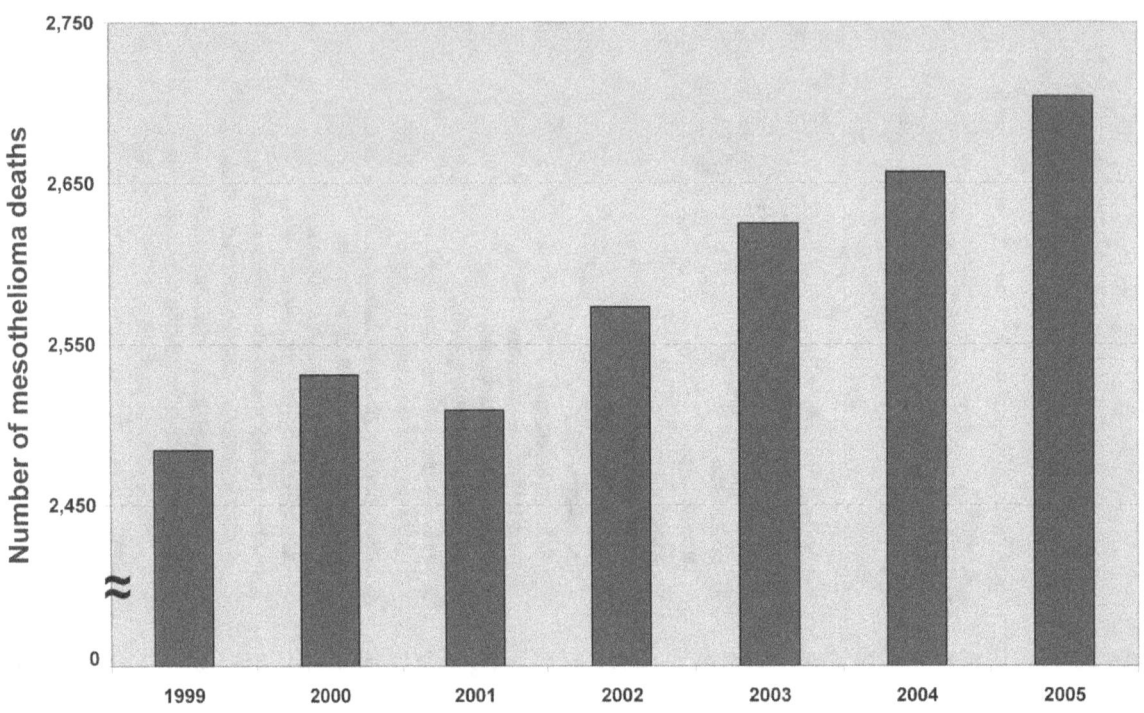

Figure 4. Number of deaths due to malignant mesothelioma, U.S. residents aged ≥15 years, 1999–2005. Source of data: NIOSH [2007b].

[Järvholm et al. 1999]. One analysis of malignant mesothelioma incidence based on the National Cancer Institute's Surveillance, Epidemiology, and End Results (SEER) Program data found that an earlier steep increase in incidence had moderated and that mesothelioma incidence may have actually peaked sometime in the 1990s in SEER-covered areas [Weill et al. 2004]. In contrast to NORMS data, which represents a census of all deaths in the entire United States, the analyzed SEER data were from areas in which a total of only about 15% of the U.S. population resides.

2.5 Workers' Home Contamination

In addition to workers' exposures, their families also have the potential for exposure to asbestos [NIOSH 1995]. Families have been exposed to asbestos when workers were engaged in mining, shipbuilding, insulating, maintenance and repair of boilers and vehicles, and asbestos removal operations. Occupants of homes where asbestos workers live may be exposed while house cleaning and laundering; these activities can result in hazardous exposure for the person performing the tasks, as well as for others in the household.

As the understanding of the effects of second-hand exposure to asbestos has increased, health effects among families of asbestos-exposed workers have been identified. Most documented cases of asbestos-related disease among workers' family members have occurred in households where women were exposed during home laundering of contaminated work clothing. In addition, children have been exposed at home by playing in areas where asbestos-contaminated shoes and work clothes were located, or where asbestos-containing materials were stored [NIOSH 1995]. As a result of these exposures, family members have been found to be at increased risk of malignant mesothelioma, lung cancer, cancer of the gastrointestinal tract, asbestosis, and nonmalignant pleural abnormalities [NIOSH 1995].

Because of the long latency periods between exposure and manifestation of asbestos-related disease, identification and intervention are difficult. Home contamination may pose a serious public health problem; however, the extent to which these health effects occur is not fully known because there are no information systems to track them [NIOSH 1995].

2.6 Clinical Issues

A thorough review of how asbestos-related diseases are diagnosed is beyond the scope of this document, and authoritative guidance on the diagnosis and attribution of asbestos-caused diseases has been published elsewhere [Anonymous 1997; British Thoracic Society Standards of Care Committee 2001; Henderson et al. 2004; ATS 2004].

The diagnosis of asbestos-caused malignancies (e.g., lung cancer and malignant mesothelioma) is almost always based on characteristic histology (or abnormal cytology in some cases). Despite research on other possible etiologies, genetic susceptibilities, and hypothesized cofactors such as simian virus 40, it is generally accepted that most cases of malignant mesothelioma are caused by exposure to asbestos or other mineral (e.g., erionite) fibers [Robinson and Lake 2005; Carbone and Bedrossian 2006]. Of particular concern to patients diagnosed with malignant mesothelioma, as well as to individuals who remain at risk due to past exposures, the disease currently is essentially incurable [British Thoracic Society Standards of Care Committee 2001]. Diagnosis may be relatively straightforward but can be difficult because of a challenging differential diagnosis [Lee et al. 2002]. Advances have been made to improve diagnostic testing for malignant mesothelioma with use of immunochemical markers and other more sophisticated histopathological analyses, and additional research is aimed at improving treatment of the disease [Robinson and Lake 2005]. Notable recent research efforts have been directed toward the development of biomarkers for mesothelioma that can be assessed by noninvasive means. A long-term goal of the biomarker research is to enable screening of high-risk individuals with sufficiently sensitive and specific noninvasive biomarkers to identify disease at an early stage, when therapeutic intervention might have a greater potential to slow the progression of the disease or be curative. Other goals are to use noninvasive biomarkers for monitoring the disease in patients treated for mesothelioma and for diagnosing the disease. Noninvasive biomarkers, including osteopontin and soluble mesothelin-related peptide, have been and continue to be evaluated, but none are considered ready for routine clinical application [Cullen 2005; Scherpereel and Lee 2007].

Nonmalignant asbestos-related diseases are diagnosed by considering three major necessary criteria: (1) evidence of structural change consistent with asbestos-caused effect (e.g., abnormality on chest image and/or tissue histology); (2) evidence of exposure to asbestos (e.g., history of occupational or environmental exposure with appropriate latency, and/or higher than normal numbers of asbestos bodies identified in lung tissue, sputum, or bronchoalveolar lavage; and/or other concurrent marker of asbestos exposure, such as pleural plaques as evidence of exposure when diagnosing asbestosis); and (3) exclusion of alternative diagnoses [ATS 2004]. The specificity of an asbestosis diagnosis increases as the number of consistent clinical abnormalities increases [ATS 2004]. In practice, only a small proportion of cases are diagnosed on the basis of tissue histopathology, as lung biopsy is an invasive procedure with inherent risks for the patient. Thus, following reasonable efforts to exclude other possible diagnoses, the diagnosis of asbestosis usually rests on chest imaging abnormalities that are consistent with asbestosis in an individual judged to have sufficient exposure and latency since first exposure.

Chest radiography remains the most commonly used imaging method for screening exposed individuals for asbestosis and for evaluating symptomatic patients. Nevertheless, as with any screening tool, the predictive value of a positive chest radiograph alone depends upon the underlying prevalence of asbestosis in the screened population [Ross 2003]. A widely accepted system for classifying radiographic abnormalities of the pneumoconioses was initially intended primarily for epidemiological use but has long been widely used for other purposes (e.g., to determine eligibility for compensation and for medicolegal purposes) [ILO 2002]. A NIOSH-administered "B Reader" Program trains and tests physicians for proficiency in the application of this system [NIOSH 2007c]. Some problems with the use of chest radiography for pneumoconioses have long been recognized [Wagner et al. 1993] and recent abuses have garnered substantial attention [Miller 2007]. In response, NIOSH recently published guidance for B Readers [NIOSH 2007d] and for the use of B Readers and ILO classifications in various settings [NIOSH 2007e].

In developed countries, conventional film radiography is rapidly giving way to digital radiography, and work is currently under way to develop digital standards and validate their use in classifying digital chest radiographs under the ILO system [Franzblau et al. 2009; NIOSH 2008a]. Progress on developing technical standards for digital radiography done for pneumoconiosis and ILO classification is under way [NIOSH 2008a]. In a validation study involving 107 subjects with a range of chest parenchymal and pleural abnormalities typical of dust-induced diseases, Franzblau et al. [2009] compared ILO classifications based on digital radiographic images and corresponding conventional chest X-ray films. The investigators found no difference in classification of small parenchymal opacities. Minor differences were observed in the classification of large parenchymal opacities, though more substantial differences were observed in the classification of pleural abnormalities typical of asbestos exposure [Franzblau et al. 2009].

Computerized tomography, especially high-resolution computed tomography (HRCT), has proven more sensitive and more specific than chest radiography for the diagnosis of asbestosis and is frequently used to help rule out other conditions [DeVuyst and Gevenois 2002]. Standardized systems for classifying pneumoconiotic abnormalities have been proposed for computed tomography but have not yet been widely adopted [Kraus et al. 1996; Huuskonen et al. 2001].

In addition to documenting structural tissue changes consistent with asbestos-caused disease, usually assessed radiographically as discussed above, the diagnosis of asbestosis relies on documentation of exposure [ATS 2004]. In clinical practice, exposure is most often ascertained by the diagnosing physician from an occupational and environmental history, assessed with respect to intensity and duration. Such a history enables a judgment about whether the observed clinical abnormalities can be reasonably attributed to past asbestos exposure, recognizing that severity of lung fibrosis is related to dose and latency [ATS 2004]. The presence of otherwise unexplained characteristic pleural plaques, especially if calcified, can also be used as evidence of past asbestos exposure [ATS 2004]. As explained by the ATS [2004], "the specificity of the diagnosis of asbestosis increases with the number of consistent findings on chest film, the number of clinical features present (e.g., symptoms, signs, and pulmonary function changes), and the significance and strength of the history of exposure." In a small minority of cases, particularly when the exposure history is uncertain or vague or when additional clinical assessment is required to resolve a challenging differential diagnosis, past asbestos exposure is documented through mineralogical analysis of sputum, bronchoalveolar lavage fluid, or lung tissue. Light microscopy can be used to detect and count asbestos bodies (i.e., asbestos fibers that have become coated with iron-containing hemosiderin during residence in the body, more generically referred to as ferruginous bodies) in clinical samples. Electron microscopy (EM) can be used to detect and count uncoated asbestos fibers in clinical samples. Methods for such clinical mineralogical analyses often vary, and valid background levels are difficult to establish. The absence of asbestos bodies cannot be used to rule out past exposure with certainty, particularly chrysotile exposure, because chrysotile fibers are known to be less persistent in the lungs than amphibole asbestos fibers and are less likely to produce asbestos bodies [De Vuyst et al. 1998; ATS 2004].

2.7 The NIOSH Recommendation for Occupational Exposure to Asbestos

NIOSH has determined that exposure to asbestos fibers causes cancer and asbestosis in humans and recommends that exposures be reduced to the lowest feasible concentration. NIOSH has designated asbestos to be a "Potential Occupational Carcinogen"§. Currently, the designation "Potential Occupational Carcinogen" is based on the classification system adopted by OSHA in the 1980s and is the only designation NIOSH uses for occupational carcinogens. After initially setting an REL at 2 asbestos fibers per cubic meter of air (f/cm³) in 1972, NIOSH later reduced its REL to 0.1 f/cm³, measured as an 8-hour time-weighted average (TWA)⁋ [NIOSH 1976]. The REL was set at a

---

§NIOSH's use of the term "Potential Occupational Carcinogen" dates to the OSHA classification outlined in 29 CFR 1990.103, and, unlike other agencies, is the only classification for carcinogens that NIOSH uses. See Section 6.1 for the definition of "Potential Occupational Carcinogen." The National Toxicology Program [NTP 2005], of which NIOSH is a member, has determined that asbestos and all commercial forms of asbestos are known to be human carcinogens based on sufficient evidence of carcinogenicity in humans. The International Agency for Research on Cancer (IARC) concluded that there was sufficient evidence for the carcinogenicity of asbestos in humans [IARC 1987b].

⁋The averaging time for the REL was later changed to 100 minutes in accordance with NIOSH Analytical Method #7400 [NIOSH 1994a]. This change in sampling time was first mentioned in comments and testimony presented by NIOSH to OSHA [NIOSH 1990a,b] and was reaffirmed in comments to MSHA in 2002, with the explanation that the 100-minute averaging time

level intended to (1) protect against the noncarcinogenic effects of asbestos; (2) materially reduce the risk of asbestos-induced cancer (only a ban can ensure protection against carcinogenic effects of asbestos); and (3) be measured by techniques that are valid, reproducible, and available to industry and official agencies [NIOSH 1976]. This REL was set at the limit of quantification (LOQ) for the phase contrast microscopy (PCM) analytical method for a 400-L sample, but risk estimates indicated that exposure at 0.1 f/cm^3 throughout a working lifetime would be associated with a residual risk for lung cancer. A risk-free level of exposure to airborne asbestos fibers has not been established [NIOSH 1976, 1984].

2.7.1 The NIOSH REL as Revised in 1990

In 1990, NIOSH [1990b] revised its REL, retaining the 0.1 f/cm^3 limit but explicitly encompassing EMPs from the nonasbestiform analogs of the asbestos minerals:

> NIOSH has attempted to incorporate the appropriate mineralogic nomenclature in its recommended standard for asbestos and recommends the following to be adopted for regulating exposures to asbestos:
>
> The current NIOSH asbestos recommended exposure limit is 100,000 fibers greater than 5 micrometers in length per cubic meter of air, as determined in a sample collected over any 100-minute period at a flow rate of 4 L/min using NIOSH Method 7400, or equivalent. In those cases when mixed fiber types occur in the same environment,

> then Method 7400 can be supplemented with electron microscopy, using electron diffraction and microchemical analyses to improve specificity of the fiber determination. NIOSH Method 7402 ... provides a qualitative technique for assisting in the asbestos fiber determinations. Using these NIOSH microscopic methods, or equivalent, airborne asbestos fibers are defined, by reference, as those particles having (1) an aspect ratio of 3 to 1 or greater; and (2) the mineralogic charz elemental composition) of the asbestos minerals and their nonasbestiform analogs. The asbestos minerals are defined as chrysotile, crocidolite, amosite (cummingtonite-grunerite), anthophyllite, tremolite, and actinolite. In addition, airborne cleavage fragments from the nonasbestiform habits of the serpentine minerals antigorite and lizardite, and the amphibole minerals contained in the series cummingtonite-grunerite, tremolite-ferroactinolite, and glaucophane-riebeckite shall also be counted as fibers provided they meet the criteria for a fiber when viewed microscopically.

The NIOSH REL for asbestos has been described in NIOSH publications and in formal comments and testimony submitted to the Department of Labor. The recommendation was based on the Institute's understanding in 1990 of potential hazards, the ability of the analytical methods to distinguish and count fibers, and the prevailing mineral definitions used to describe covered minerals.

2.7.1.1 Rationale for Encompassing Nonasbestiform Analogs of Asbestos Varieties Within the REL

NIOSH's rationale for recommending that nonasbestiform analogs of the asbestos minerals be encompassed within the policy definition

---

would help "to identify and control sporadic exposures to asbestos and contribute to the overall reduction of exposure throughout the workshift" [NIOSH 2002].

of airborne asbestos fibers was first articulated in comments and testimony to OSHA [NIOSH 1990a,b]. According to the 1990 testimony, NIOSH based its recommendation on three elements:

- The first element comprised results of epidemiological studies of worker populations exposed to EMPs from nonasbestiform mineral analogs of the asbestos varieties (e.g., cleavage fragments). The 1990 testimony characterized the existing evidence as equivocal for excess lung cancer risk attributable to exposure to such nonasbestiform EMPs.

- The second element comprised results of animal carcinogenicity studies involving experimental intrapleural or intraperitoneal administration of various mineral particles. The 1990 testimony characterized the results of the studies as providing strong evidence that carcinogenic potential depends on a mineral particle's length and width and reasonable evidence that neither chemical composition nor mineralogical origin is a critical factor in determining a mineral particle's carcinogenic potential.

- The third element comprised the lack of routine analytical methods to accurately and consistently distinguish between asbestos fibers and nonasbestiform EMPs in samples of airborne fibers. The 1990 testimony argued that asbestiform and nonasbestiform minerals can occur in the same area and that determining the location and identification of tremolite asbestos, actinolite asbestos, and anthophyllite asbestos within deposits of their nonasbestiform mineral analogs can be difficult, resulting in mixed exposures for some mining operations and downstream users of their mined commodities.

Given the inconclusive epidemiological evidence for lung cancer risk associated with exposure to cleavage fragments (first bullet, above), NIOSH took a precautionary approach and relied upon the other two elements to recommend that the 0.1 f/cm^3 REL for airborne asbestos fibers also encompass EMPs from the nonasbestiform analogs of the asbestos minerals. In fact, the 1990 NIOSH testimony included an explicit assertion that the potential risk of lung cancer from exposure to EMPs (of the nonasbestiform asbestos analog minerals) warranted limiting such exposures. However, even if such EMPs were not hazardous, the inability of analytical methods to accurately distinguish countable particles as either asbestos fibers or cleavage fragments (of the nonasbestiform analog minerals) presents a problem in the context of potentially mixed exposures (i.e., asbestos fibers together with EMPs from the nonasbestiform analogs). NIOSH's 1990 recommendation provided a prudent approach to mixed dust environments with potential exposure to asbestos; limiting the concentration of all countable particles that could be asbestos fibers to below the REL would ensure that the asbestos fiber component of that exposure would not exceed the REL.

Some scientists and others have questioned NIOSH's rationale for including EMPs from nonasbestiform amphibole minerals in its definition of "airborne asbestos fibers." Mineralogists argue that these EMPs do not have the morphological characteristics required to meet the mineralogical definition of "fibers"; acicular and prismatic amphibole crystals and cleavage fragments generated from the massive habits of the nonasbestiform analogs of the asbestos minerals are not true mineralogical fibers. Others have opined that the scientific literature does not demonstrate any clear health risks associated with exposure to the nonasbestiform EMPs covered by the NIOSH "airborne asbestos fiber" definition.

Whether or not to include EMPs from nonasbestiform analogs of the asbestos minerals in federal regulatory asbestos policies has been the subject of long-standing debate. The exposure-related toxicity and health effects associated with the various morphologies (e.g., acicular, prismatic) of the nonasbestiform analogs of the asbestos minerals continue to be a central point in the debate. In 1986, OSHA revised its asbestos standard and included nonasbestiform anthophyllite, tremolite, and actinolite (ATA) as covered minerals within the scope of the revised standard [OSHA 1986]. OSHA's decision to include nonasbestiform ATA proved controversial. In a 1990 proposal to reverse this revision, OSHA [1990] noted that there were "a number of studies which raise serious questions about the potential health hazard from occupational exposure to nonasbestiform tremolite, anthophyllite and actinolite," but that the "current evidence is not sufficiently adequate for OSHA to conclude that these mineral types pose a health risk similar in magnitude or type to asbestos."

In the preamble to the final rule removing nonasbestiform ATA from its asbestos standard, OSHA [1992] stated that:

> ...various uncertainties in the data** and a body of data showing no carcinogenic effect, do not allow the Agency to perform qualitative or quantitative risk assessments concerning occupational exposures. Further, the subpopulations of nonasbestiform ATA which, based on mechanistic and toxicological data, may be associated with a carcinogenic effect, do not appear to present an occupational risk. Their presence in the workplace is not apparent from the record evidence.

In its 2005 proposed rule for asbestos, MSHA stated that substantive changes to its asbestos definition were beyond the scope of the proposed rule and chose to retain its definition of asbestos, which "does not include nonfibrous or nonasbestiform minerals" [MSHA 2005]. These decisions are reflected in MSHA's final rule published in 2008 [MSHA 2008]. In formal comments during the rulemaking process, NIOSH agreed with MSHA's decision not to modify its asbestos definition in the current rulemaking, stating that "NIOSH is presently re-evaluating its definition of asbestos and nonasbestiform minerals, and will work with other agencies to assure consistency to the extent possible" [NIOSH 2005]. In the interim, the following subsections provide more detail on the three elements comprising NIOSH's rationale for recommending in 1990 that nonasbestiform analogs of the asbestos minerals be encompassed by the REL.

2.7.1.1.1 Epidemiological Studies

The first element of NIOSH's rationale for recommending in 1990 that nonasbestiform analogs of the asbestos minerals be encompassed by the REL related to evidence from epidemiological studies.

A number of epidemiological studies have been conducted to evaluate the health risk to workers reported to be exposed to EMPs. Epidemiological studies have been conducted in the Gouverneur talc district of upstate New York in

---

**OSHA was referring to the scientific data on which NIOSH based its own carcinogenic health effect recommendation to OSHA.

which the deposits have been reported to contain both amphibole cleavage fragments and transitional mineral fibers [Beard et al. 2001]. Differing interpretation of the analysis of these particles has caused considerable disagreement over the asbestos content of these talc deposits [Van Gosen et al. 2004]. Because of the disagreement over the classification of these mineral particles, the epidemiological studies of talc miners and millers in upstate New York are reviewed here. Additional epidemiological studies of worker populations with exposures to EMPs are also reviewed, including studies conducted at the Homestake gold mine in South Dakota and the taconite mining region of northeastern Minnesota. The findings from these investigations are reviewed in detail below.

Studies of New York Talc Miners and Millers

Workers exposed to talc[††] have long been recognized to have an increased risk of developing pulmonary fibrosis, often referred to as talc pneumoconiosis [Siegel et al. 1943; Kleinfeld et al. 1955]. Talc-exposed workers have also been reported to have an increased prevalence of pleural abnormalities, including pleural thickening and pleural plaques [Siegel et al. 1943; Gamble et al. 1982].

Several more recent epidemiological studies and reviews have been conducted of workers employed in talc mines and mills in upstate New York [Brown et al. 1979, 1990; Kleinfeld et al. 1967, 1974; Lamm and Starr 1988; Lamm et al. 1988; Stille and Tabershaw 1982; Gamble 1993; Honda et al. 2002; Gamble and Gibbs 2008].

[††]The characteristics of talc deposits vary with location [Van Gosen et al. 2004]. Unless otherwise specified, the talc referred to in this section is restricted to the deposits in New York State which contain talc and other minerals.

Excessive rates of mesothelioma have been reported for Jefferson County, which is immediately adjacent to the Balmat, Fowler, and Edwards talc-mining areas of St. Lawrence County in New York [Vianna et al. 1981; Enterline and Henderson 1987; Hull et al. 2002]. In a study of all histologically confirmed mesothelioma cases reported to New York State's tumor registry from 1973 to 1978, Vianna et al. [1981] reported six cases from Jefferson County, resulting in a mesothelioma rate for that county more than twice that of New York State (excluding New York City). In a national study of pleural cancer mortality (ICD-9 163) from 1966 through 1981, Enterline and Henderson [1987] reported 4 mesothelioma cases in Jefferson County females (0.6 expected) and 7 cases in Jefferson County males (1.4 expected), resulting in county mesothelioma rates that were the second and sixth highest county-specific rates in the nation for females and males, respectively (both $p < 0.01$). More recently, Hull et al. [2002] updated the Enterline and Henderson pleural cancer mortality analysis for Jefferson County, reporting 5 new male cases (2 expected) and 3 new female cases (0.5 expected) through 1997 and describing Jefferson County mesothelioma death rates as "5–10 times the background rate." A potential limitation of the Enterline and Henderson [1987] and Hull et al. [2002] analyses is that they relied on ICD code 163 ("malignant neoplasms of the pleura, mediastinum, and unspecified sites") as a surrogate identification for malignant mesothelioma. That code lacked specificity and sensitivity for mesothelioma; in a study of Massachusetts deaths, many non-mesothelioma malignancies involving the pleura were assigned code 163 and most mesotheliomas were not assigned code 163 [Davis et al. 1992]. The more recent ICD-10 system, which has been used since 1999 to code death certificate data in the United

States, includes a discrete code for malignant mesothelioma. Based on that new ICD-10 code, the age-adjusted death rates (per million population) for 1999–2004 were 12.9 (based on 5 mesothelioma deaths) for Jefferson County and 10.9 (based on 5 mesothelioma deaths) for St. Lawrence County. These are similar to the overall U.S. mesothelioma death rates for this same period (based on a total of 15,379 mesothelioma deaths) of 11.5 per million [NIOSH 2007b].

An excess of lung cancer has also been reported following several epidemiological studies of New York talc mines and mills [Kleinfeld et al. 1967, 1974; Brown et al. 1990; Lamm and Starr 1988; Stille and Tabershaw 1982; Lamm et al. 1988; Honda et al. 2002]. The most extensive research has been conducted on workers at the talc mine and mills owned by RT Vanderbilt Company, Inc. (RTV), located in St. Lawrence County. A significant excess of mortality from nonmalignant respiratory disease (NMRD) has been consistently noted in these studies. These studies have also generally demonstrated an approximately twofold to threefold increase in lung cancer mortality among these workers [Brown et al. 1990; Honda et al. 2002; Lamm et al. 1988]. The lung cancer excess has been reported to be particularly high among workers with more than 20 years since their first exposure (latency), which is a pattern consistent with an occupational etiology [Brown et al. 1979, 1990]. Authors of several studies have questioned whether the excess of lung cancer observed in these studies is due to employment at the RTV mines and mills or to other factors [Honda et al. 2002; Lamm et al. 1988; Stille and Tabershaw 1982]. Attributing these findings to employment in the RTV mine is difficult because there were numerous mines operating in upstate New York, and the mineralogical composition of the ores varied substantially [Petersen et al. 1993]. A high smoking rate among the workers at the RTV mine and mills has been suggested as one possible explanation for the excess lung cancer mortality [Kelse 2005; Gamble 1993]. However, it is generally considered implausible that confounding by smoking in occupational cohort studies could explain such a large (i.e., about twofold to threefold) increase in lung cancer mortality [Steenland et al. 1984; Axelson and Steenland 1988; Axelson 1989].

The most persuasive argument against a causal interpretation of these findings is that the lung cancer excess in this study population did not increase with duration and measures of exposure to dust [Lamm et al. 1988; Stille and Tabershaw 1982; Honda et al. 2002]. Also, the excess of lung cancer in this cohort has been reported to be limited to workers with short employment (<1 year) [Lamm et al. 1988] and to workers who have been employed in other industries where exposure to carcinogens may have occurred prior to their working in the RTV mine and mills [Lamm et al. 1988; Stille and Tabershaw 1982]. The latter observation could be explained by there simply being too few workers and inadequate follow-up of workers who have worked only at RTV to provide the statistical power necessary to demonstrate an increased lung cancer risk. For example, in one of the studies only 10% of the decedents were reported to have not worked in other industries prior to their employment at RTV [Stille and Tabershaw 1982].

In the most recent study of RTV miners and Millers, Honda et al. [2002] examined lung cancer mortality in relation to quantitative estimates of exposure to respirable talc dust [Oestenstad et al. 2002]. As in previous studies, mortality from lung cancer was found to be significantly elevated (standardized mortality ratio [SMR] = 2.3; 95% confidence interval [95% CI] = 1.6–3.3), with the most pronounced excess of lung cancer mortality found in short-term workers (<5 years). An exposure-response relationship was observed in the study between cumulative exposure to respirable dust and NMRD and pulmonary fibrosis.

One explanation that has been offered for the lack of exposure-response in these studies is that the observed excess of lung cancer was a result of exposures from employment prior to starting work at RTV. It has been suggested that many of these workers may have had prior employment in neighboring talc mines in upstate New York with similar exposures to talc [NIOSH 1980]. Not considering exposures at these other mines could have substantially impacted results of exposure-response analyses. Exposures to dust may also have been substantially higher in the neighboring mines than in the RTV mine [Kelse 2005]. Because RTV workers may have had exposures to dust in other mines, their exposures may have been underestimated, which could explain the observed lack of an exposure-response relationship in the epidemiological studies of RTV workers. There is also evidence to suggest that RTV workers may have been exposed to lung carcinogens from prior work in non-talc industries [Lamm et al. 1988].

Gamble [1993] conducted a nested case-control study of lung cancer in the RTV cohort to further explore whether factors unrelated to exposures at RTV, such as smoking and exposures from prior employment, might be responsible for the observed excess of lung cancer among RTV workers. Cases and controls were identified from among 710 workers who were employed between 1947 and 1958, and vital status was ascertained through 1983. All individuals whose underlying cause of death was lung cancer were included as cases (n = 22).

Three controls (n = 66) for each case were selected from members of the cohort who had not died of NMRD or accidents, and they were matched to cases on the basis of dates of birth and hire. Controls were also required to have survived for as long as their matched case. Information on smoking and work histories was obtained by interviewing the case (if alive) or relatives. An attempt was made to verify information on previous employment by checking personnel records and by contacting previous employers. A panel of epidemiologists and industrial hygienists classified previous non-talc employment with regard to the probability of occupational exposure to a lung cancer risk.

As in previous investigations of the RTV cohort, Gamble [1993] found that the risk of lung cancer decreased with increasing duration of employment at RTV. This was true among both smokers and nonsmokers, and also when individuals with inadequate time since first exposure (<20 years) and short duration of employment were excluded. Lung cancer risk was also found to decrease with increasing probability of exposure to lung carcinogens from non-talc employment. A positive exposure-response relationship was evident when non-RTV talc exposures were included in the analysis, although this relationship was not statistically significant.

This study by Gamble [1993] does not provide support for the hypothesis that prior employment in non-talc industries was responsible for the excess of lung cancer observed among RTV workers. The author interpreted his findings as providing support for the argument that the excess of lung cancer was due to confounding by smoking, on the basis of the fact that smoking was strongly associated with lung cancer risk and on the observation that the exposure-response relationship with talc was more strongly negative (inverse) in analyses restricted to smokers than among all study subjects. However, it is no surprise that an association was observed between smoking and lung cancer, and the fact that the negative (inverse) exposure-response trend was stronger among smokers does not explain why the cohort as a whole experienced much higher lung cancer rates than expected.

Only two cases of pleural mesothelioma have been reported in the cohort studies of RTV miners and millers [Honda et al. 2002], and those authors reported that it was unclear whether these cases are attributable to exposure to talc at the RTV mine and mills. One of the individuals had worked for only a short time in a job with minimal talc exposure, had previously worked for many years in the construction of a talc mine, and had subsequently worked on repairing oil heating systems. The other case developed only 15 years after first exposure to RTV talc, although mesothelioma has more typically been observed to develop after much longer times from first exposure. NIOSH has recently received unpublished reports of additional cases of pleural mesothelioma among workers at RTV or its predecessor, International Talc [Abrams 2010; Maimon 2010], and a report of at least one worker using products from the mine who may also have died from mesothelioma [Satterley 2010]. NIOSH has also received unpublished observations concerning findings from chest radiographs RTV has obtained on its workers semiannually since 1985. Although pleural plaques were more frequently observed, the program has identified only one worker with irregular parenchymal opacities consistent with asbestosis-like disease, results that a consulting pulmonologist "did not find … to be consistent with that of a workforce exposed to asbestos" [Kelse et al. 2008].

Van Gosen [2007] has reported that fibrous varieties of talc, tremolite, and anthophyllite occur in the tremolite-talc deposits of the Governeur talc mining district of upstate New York. It has been debated whether the fibrous amphiboles in those talc ores meet the criteria of asbestos, with the debate centered on the complex and unusual transitional fibers composed partly of talc and partly of anthophyllite. NIOSH [1980] and others [Van Gosen 2006; Webber 2004] have reported that dust from the RTV mine, which is in the Governeur talc mining district, contains asbestos. In an industrial hygiene assessment conducted at RTV mines by NIOSH [1980], X-ray diffraction and petrographic microscopic analyses of talc product samples found them to contain 4.5%–15% anthophyllite (some of which was categorized as asbestos). In contrast, Kelse [2005] reported the percentage by weight of talc from the RTV mine in upstate New York as 1%–5% nonasbestiform anthophyllite. In airborne samples collected by NIOSH at the mine and mill and analyzed by TEM, 65% of the EMPs that were longer than 5 μm were anthophyllite and 7% were tremolite, and much of the tremolite was from a nonfibrous habit [NIOSH 1980]. Kelse [2005] reported that up to 1.8% was from an asbestiform habit, though the asbestiform component was reported not to be asbestos. Serpentine and amphibole minerals typically develop through the alteration of other minerals. Consequently, they may exist as partially altered minerals having variations in elemental compositions. Minerals undergoing this alteration are often called "transitional minerals." Thus the elemental composition of individual mineral particles can vary within a mineral deposit containing transitional minerals, which could account for differences in the reported composition of talc from the RTV mine.

A limitation of the epidemiological studies of RTV talc workers is the lack of an exposure-response analysis based on direct measurements of airborne EMP concentrations. Most of the studies used tenure as a surrogate for exposure, and the exposure metric used in the Honda et al. [2002] study was respirable dust, which may not be correlated with exposure to EMPs. Relationships between health outcomes and exposure to an agent of interest can be attenuated when a nonspecific exposure indicator is used as a surrogate for exposure to the agent of interest [Blair et al. 2007; Friesen et al. 2007]. Thus, when the exposure index used to assess the effect of EMPs is based on a surrogate measure, such as respirable dust, rather than on specific measurement of EMP concentrations, the lack of an exposure-response relationship between the exposure index and the health outcome might not accurately reflect the true risk, particularly where the composition of a mixed dust exposure varies by work area.

Finally, a cohort study of Vermont talc miners and millers has some relevance for interpreting the findings from the studies of New York talc workers [Selevan et al. 1979]. Airborne dust samples and talc bulk samples from all locations in the mines and mill studied were analyzed by X-ray diffraction or analytical EM, which showed the talcs to be similar in composition; no asbestos was detected in any of the samples. To determine potential differences between the talcs mined and milled by a company no longer in operation, seven samples of talcs were obtained and also analyzed by X-ray diffraction or analytical EM. These samples were reported to be free of asbestos contamination and the mineral composition was within the range of samples from then-producing operations. The one remaining company, no longer operating at that time, had been located in the same geographic area as two of the then-producing companies. The talc in one of the closed mines was reported to have had "cobblestones" of serpentine rock that were highly tremolitic. In the mining of this talc, these cobblestones were reported to have been avoided and/or discarded. Therefore, miners may have been exposed to tremolite, but it is unlikely that the millers from this company were exposed [Selevan 1979]. Van Gosen [2004] described the talc deposits in this area as "associated spatially with serpentinite masses that in some areas host well-developed chrysotile asbestos [Bain 1942; Cady et al. 1963]. The alteration zone locally contains actinolite, tremolite, anthophyllite, and (or) cummingtonite, as described by Cady et al. [1963]. Zodac [1940] described 'radiating masses of fibrous actinolite, which often have to be handled carefully as the needlelike crystals may penetrate fingers, are common on the dumps' in a talc quarry near Chester, Vermont. The amphiboles present were not identified as asbestiform.

A statistically significant excess of NMRD mortality was observed among the millers (SMR = 4.1; 95% CI = 1.6–8.4), but not among the miners (SMR = 1.6; 95% CI = 0.20–9.6), in this study [Selevan 1979]. In contrast, respiratory cancer mortality was found to be significantly elevated among the miners (SMR = 4.3; 95% CI = 1.4–10), but not among the millers (SMR = 1.0; 95% CI = 0.12–4.0). The authors suggested that their respiratory cancer findings might be due to non-talc exposures, such as radon progeny, because exposures to talc dust were higher among millers than miners. The pattern of excess respiratory cancer observed in this study is similar to that reported in studies of RTV miners and millers. It has been noted [Lamm and Starr 1988] that this provides evidence against the hypothesis that the lung cancer excess among RTV miners is caused by exposure to the EMPs in RTV talc, because these were not identified in Vermont talc.

In summary, excesses of pulmonary fibrosis and pleural plaques are generally recognized to have occurred among workers exposed to talc. Pleural cancer mortality rates have been reported to be significantly elevated in Jefferson County, adjacent to St. Lawrence County, where the New York talc industry has been located. However, recent death certificate data for 1999–2004 do not suggest a particularly high rate of mesothelioma in St. Lawrence County or Jefferson County. Also, aspects of the few cases of mesothelioma that have been carefully evaluated in the published studies of New York talc miners make it unclear whether the cases are attributable to employment in the talc industry. Lung cancer mortality has been consistently reported to be elevated in studies of New York talc miners. However, attribution of this excess to exposures to dust containing talc and other minerals has been questioned, because the lung cancer excess was generally found to be most pronounced in short-term workers and did not increase with cumulative exposure to talc dust. It is highly unlikely that chance or confounding from smoking or prior mining exposures fully accounts for the lung cancer excess observed in these studies. These findings may be at least partly explained by employment in other industries, including other mines in upstate New York.

Studies of Homestake Gold Miners

Three groups of investigators have conducted retrospective cohort studies of miners at the Homestake gold mine in South Dakota, with somewhat different and overlapping cohort definitions. Gillam et al. [1976] studied 440 white males who were employed as of 1960 and who had worked underground for at least 5 years in the mine. McDonald et al. [1978] conducted a retrospective cohort study of 1,321 men who had retired and worked for at least 21 years in the mine as of 1973 and were followed for vital status until 1974. Brown et al. [1986] conducted a retrospective cohort study of 3,328 miners who had worked for at least 1 year between 1940 and 1965, with follow-up of vital status to 1977. Follow-up of this same cohort was subsequently updated to 1990 by Steenland and Brown [1995]. Exposures of potential concern at this mine include crystalline silica, radon progeny, arsenic, and nonasbestiform EMPs. The longer (>5 µm) nonasbestiform EMPs have been reported to be primarily cummingtonite-grunerite (69%), but tremolite-actinolite (15%) and other nonasbestiform amphibole varieties (16%) were also detected [Zumwalde et al. 1981]. Most of the EMPs observed by TEM (70%–80%) were shorter than 5 µm; for the entire population of EMPs, the geometric mean length was 3.2 µm and the geometric mean diameter was 0.4 µm.

There is very little evidence of an excess of mesothelioma in the studies of Homestake gold miners. One case of mesothelioma with "low" dust exposure was noted in the study by McDonald et al. [1978]. Slight excesses of cancer of the peritoneum (4 cases; SMR = 2.8; 95% CI = 0.76–7.2) and other respiratory cancer (3 cases; SMR = 2.5; 95% CI = 0.52–7.4) were described in the most recently reported study [Steenland and Brown 1995]. These categories might be expected to include cases of mesothelioma; however, mesothelioma was not mentioned on the death certificates for these cases.

Significant excesses in mortality from tuberculosis and pneumoconiosis (mainly silicosis) were observed in all of the studies. An excess of respiratory cancer (10 cases: SMR = 3.7; 95%

CI = 1.8–6.7) was noted in the earliest study, by Gillam et al. [1976]. Respiratory cancer mortality was not found to be elevated (34 cases; SMR = 1.0; 95% CI = 0.71–1.4), and there was only weak evidence that it increased with level of exposure, in the study by McDonald et al. [1978]. A slight excess of lung cancer (115 cases; SMR = 1.1; 95% CI = 0.94–1.4) was noted in the most recent study, in comparison with U.S. mortality rates [Steenland and Brown 1995]. This lung cancer excess was more pronounced when county rates were used as the referent (SMR = 1.3; 95% CI = 1.0–1.5), and even more so when South Dakota state rates were used (SMR = 1.6; 95% CI = 1.3–1.9). The excess was also increased (versus U.S. rates; SMR = 1.3; 95% CI = 1.0–1.6) when the analysis was restricted to individuals with at least 30 years of time since first exposure (latency). Lung cancer mortality was not found to increase with estimated cumulative exposure to dust in this study, though a clear exposure-response trend was observed for pneumoconiosis. The limited available data on smoking habits indicated that miners in this cohort smoked slightly more than the U.S. general population in a 1960 survey.

Taken together, the studies of Homestake gold miners provide, at best, weak evidence of an excess risk of lung cancer. Although small excesses of lung cancer have been reported in the most recent studies of the Homestake gold miners, the increased mortality has not been found to increase with measures of cumulative dust exposure. The uncertainty of the relationship between contemporaneous dust and EMP exposures hinders the usefulness of historical dust measurement data in estimating EMP exposures [Zumwalde et al. 1981]. Thus the lack of cancer noted in these studies is largely uninformative with respect to the hypothesis that nonasbestiform EMPs are associated with increased risk of respiratory diseases in this population.

Studies of Taconite Miners

There has been a long history of concern about a potential association between exposures associated with the taconite iron ore industry in northeastern Minnesota and the potential cancer risk. This concern started in 1973, when amphibole fibers were found in the Duluth water supply and were traced to tailings that had been disposed of in Lake Superior by the Reserve Mining Company. Extensive sampling and analysis of areas of the Peter Mitchell taconite iron ore mines were recently conducted by Ross et al. [2007], who reported finding "no asbestos fibers of any type" in the mines. However, they did find and describe fibrous ferroactinolite, fibrous ferrian sepiolite, fibrous grunerite-ferroactinolite, and fibrous actinolite in ore samples, some of which was very thin (<0.01 µm) with a very high aspect ratio. They estimated fibrous amphibole material to represent "a tiny fraction of one percent of the total rock mass of this taconite deposit" [Ross et al. 2007].

Several epidemiological studies have examined mortality of miners working in the taconite mines and mills of Minnesota. Higgins et al. [1983] published the earliest study, which examined the mortality of approximately 5,700 workers employed at the Reserve Mining Company between 1952 and 1976 and followed up to 1976. Overall mortality (SMR = 0.87) and mortality from respiratory cancer (15 cases; SMR = 0.84) were both less than expected. Respiratory cancer mortality was not found to be increased among workers with at least 15 years since first exposure (latency) and did not increase with estimated cumulative exposure to dust. The maximum follow-up of this cohort was 24 years, which is probably too short to be able to detect increased mortality from lung cancer or mesothelioma.

Cooper et al. [1988, 1992] have reported on the mortality experience of 3,431 miners and millers who were employed in the Erie or Minntac mines and mills for at least 3 months between 1947 and 1958. Follow-up of the cohort, initially to 1983 [Cooper et al. 1988], was extended to 1988 in their more recent update [Cooper et al. 1992]. Comparisons were made with white male mortality rates for Minnesota and for the U.S. population. Mortality from respiratory cancer was found to be slightly less than expected in this study (106 cases, based on Minnesota rates: SMR = 0.92; 95% CI = 0.75–1.1). Respiratory cancer mortality was close to the expected value (46 cases, on the basis of Minnesota rates: SMR = 0.99; 95% CI = 0.72–1.3) among workers with more than 20 years since first exposure (latency).

A statistically significant excess of mesothelioma has been reported in northeastern Minnesota, the area in which the taconite mining and milling industry is located [MDH 2007]. In its most recent report, the Minnesota Department of Health (MDH) reported that a total of 159 cases occurred in this region during the period of 1988 to 2006. The mesothelioma rate for males was approximately twice the expected rate, based on the rest of the state (146 cases: rate ratio [RR] = 2.1; 95% CI = 1.8–2.5), whereas the rate for females was less than expected (RR = 0.72; 95% CI = 0.38–1.2). The fact that the excess of mesothelioma was observed only among males strongly suggests an occupational etiology. In addition to the taconite industry, a plant producing asbestos ceiling tiles (Conwed Corporation) was located in the northeastern Minnesota region. From 1958 to 1965 amosite was used at Conwed, and from 1966 to 1974 chrysotile was used [Mandel 2008]. The MDH has initiated epidemiological studies of mesothelioma incidence among workers at the Conwed Corporation and at the iron mines in northeastern Minnesota. The records from a cohort of approximately 72,000 iron miners and from 5,700 Conwed workers have been linked with a mesothelioma data registry. Between 1988 and 2007, a total of 58 mesothelioma cases among the miners and 25 cases among the Conwed workers were identified. Because only 3 of the 58 mesothelioma cases identified in the miner cohort had also been employed at Conwed, it is unlikely that the mesothelioma excess in miners could be explained by asbestos exposures during employment at the Conwed ceiling tile facility [MDH 2007].

Brunner et al. [2008] have recently reported findings from an MDH study of mesothelioma cases occurring among iron miners between 1988 and 1996. The job histories of the cases were reviewed for evidence of exposure to commercial asbestos. Mining jobs were identified from company personnel files. Nonmining employment information was obtained from worker application files, worker compensation records, and obituaries. Potential asbestos exposures for jobs held in the mining industry were identified by interviewing 350 workers, representing 122 occupations and 7 different mining companies. To estimate the probability and intensity of potential exposure to commercial asbestos in each of the jobs, an expert panel rated the potential for asbestos exposure on the basis of these interviews, available job descriptions from the relevant time period, and their knowledge of the mining environment. The job histories for 15 of 17 iron miners known to have developed mesothelioma were considered sufficient for the study. Eleven of these 15 were reported to have had probable exposure, 3 were reported to have possible exposure to commercial asbestos, and 1 did not have an identifiable source of exposure to commercial asbestos. The asbestos exposures were from nonmining jobs (4 cases), mining jobs (4 cases), or both (6 cases). The findings from this study suggest that the excess of mesothelioma observed among taconite miners might be explained by exposure to commercial asbestos rather than from the nonasbestiform amphibole EMPs generated during iron ore processing. However, this was a case series and it was not possible to determine whether commercial asbestos exposure was different in the cases than in the cohort as a whole or in a control group. This study also did not include the 41 additional mesothelioma cases that have been reported by the MDH since 1996 [MDH 2007].

The age-adjusted rates of deaths due to malignant mesothelioma during the period 2000–2004 for the two counties considered to be in the iron range (Itasca and St. Louis Counties) are reported as 30.4 and 31.3 per million, respectively, which are 2.6 and 2.7 times the U.S. average [NIOSH 2009a]. Two counties (Carlton and Koochiching Counties) adjacent to the iron range counties have malignant mesothelioma rates of 55.3 and 77.5 per million, which are 4.8 and 6.7 times the U.S. average. The age-adjusted asbestosis rates during 1995–2004 were reported for only two of these counties: St. Louis County (7.9 per million, which is 1.3 times the U.S. average) and Carlton County (54.0 per million, or 8.9 times the U.S. average) [NIOSH 2009b].

In summary, the results from cohort mortality studies of taconite miners and millers in Minnesota have not provided evidence of an increased risk of respiratory cancer or mesothelioma. This appears to be somewhat in conflict with reports from the MDH that mesothelioma incidence is significantly elevated among males (but not females) in northeastern Minnesota and that a large number of these cases involved workers in the Minnesota taconite industry. There is some evidence that these cases could, at least in part, be related to exposures to commercial asbestos that occurred in or outside of the taconite mining industry, but further research on this question is needed. The MDH is currently working with researchers at the University of Minnesota School of Public Health on a mesothelioma case-control study, a respiratory morbidity study, and a mortality study of the iron miners of northeastern Minnesota [MDH 2007].

Summary of Epidemiological Studies of Cohorts Exposed to Nonasbestiform EMPs

The results from studies of populations exposed to nonasbestiform EMPs do not provide clear answers regarding the toxicity of these EMPs. A number of features limit their usefulness. First, the populations in these studies were exposed to a complex mixture of particles. Although an excess of pneumoconiosis has been observed in the studies of Homestake gold miners and New York talc workers, the extent to which these findings are attributable to their exposures to nonasbestiform EMPs cannot be determined. A potential limitation of the New York talc studies is that, if the EMPs do include fibers from asbestiform minerals, as reported in the literature [NIOSH 1980; Van Gosen 2006; Webber 2004], it is difficult to determine whether the observed health effects can be attributed to them, given the heterogeneous mineral composition of exposures.

Another major limitation of these studies is that they lack adequate information on past exposure to EMPs. An excess of respiratory cancer was observed in the studies of New York talc workers, and a small excess was observed in the most recent study of Homestake gold miners. In both studies, the excess of respiratory cancer was not found to increase with cumulative exposure to dust. Relationships between health outcomes and exposure to an agent of interest can be attenuated when a nonspecific exposure indicator is used as a surrogate for exposure to that agent [Blair et al. 2007; Friesen et al. 2007]. Thus, when the exposure index used to assess the effect of EMPs is based on a surrogate measure, such as respirable dust, rather than on specific measurement of EMP concentrations, the lack of an exposure-response relationship between the exposure index and the health outcome might not accurately reflect the true risk, particularly where the composition of a mixed exposure varies by work area. Interpretation of findings from the New York talc studies has been further complicated by the employment of the workers elsewhere, including employment at other talc mines in the area. Lack of positive findings from exposure-response analyses in the New York talc studies of RTV miners and millers could also have resulted from exposure misclassification—possible under-ascertainment of exposure to talc and other mineral particles caused by not considering exposures at neighboring talc mines or previous employment in the same mine under different ownership.

The reliability of death certificate information is another major limitation, particularly for the diagnosis of mesothelioma. Mesothelioma did not have a discrete ICD code until the 10th revision of the ICD, used for U.S. death certificate data only since 1999. This likely explains the discordance between the apparent recent lack of excess mesothelioma deaths in an upstate New York county adjacent to the county in which talc mines and mills have been located and the excess "mesothelioma" death rates previously reported in that county. This may explain the apparent contradiction between the lack of an excess of mesothelioma in the cohort studies of taconite miners, and the excess of mesothelioma that has been reported in the more recent studies based on a mesothelioma registry in northeastern Minnesota.

Finally, the lack of information on cigarette smoking habits of the studied workers is a major issue in interpreting the findings for respiratory cancer in these studies. Concern about cigarette smoking in occupational cohort studies is generally based on the assumption that blue collar workers smoke more than the general population. However, the extent of this bias is generally not expected to be able to account for more than a 50% increase in lung cancer risk and is unlikely to explain the twofold to threefold risk reported in the New York talc studies. Confounding by smoking could conceivably explain the small excess of lung cancer that has been reported in the most recent study of Homestake gold miners [Steenland and Brown 1995]. However, smoking may have introduced a negative bias in some of these studies. Cigarette smoking has been reported to have been banned in the Homestake gold mines [Brown et al. 1986] and in the underground taconite mines [Lawler et al. 1985]. Preventing workers from smoking at work could have negatively biased the lung cancer findings in these studies.

Because of the study limitations described above, the findings from these studies should best be viewed as providing inconclusive as opposed to negative evidence regarding the health effects associated with exposures to nonasbestiform EMPs. To be more informative, additional studies of these populations would need improved characterizations of exposure to EMPs, smoking status, and exposures associated with other employment. Additional studies of these populations should be pursued if these improvements are deemed feasible.

2.7.1.1.2 Animal Studies

The second element of NIOSH's rationale for recommending in 1990 that nonasbestiform analogs of the asbestos minerals be encompassed by the REL is related to evidence from animal studies. Citing several original studies and reviews [Stanton et al. 1977, 1981; Wagner et al. 1982; Muhle et al. 1987; Pott et al. 1974, 1987; Lippmann 1988], NIOSH [1990a] concluded that they provided evidence that fiber dimension (and not fiber composition) was the major determinant of carcinogenicity for mineral fibers, stating that

> *Literature reviews by Lippmann [1988] and Pott et al. [1987] enhance the hypothesis that any mineral particle can induce cancer and mesothelioma if it is sufficiently durable to be retained in the lung and if it has the appropriate aspect ratio and dimensions. Similarly, Wagner [1986] concluded that all mineral particles of a specific diameter and length size range may be associated with development of diffuse pleural and peritoneal mesotheliomas.*

That general conclusion notwithstanding, a study by Smith et al. [1979] that was not cited by NIOSH in 1990 addressed the specific question of carcinogenicity of EMPs from nonasbestiform amphiboles. Pleural tumor induction by intrapleural (IP) injection challenge in hamsters was compared for various challenge materials, including two asbestiform tremolites and two nonasbestiform (prismatic) tremolitic talcs. In contrast to the two asbestiform tremolites, which induced tumors in 22% and 42% of challenged hamsters at the higher dose, no tumors resulted following challenge with either of the two nonasbestiform tremolites [Smith et al. 1979]. In its rule-making, OSHA noted several limitations of the study, including the small number of animals studied, the early death of many animals, and the lack of systematic characterization of fiber size and aspect ratio [OSHA 1992]. One of the nonasbestiform tremolitic talcs was later analyzed and determined to have tremolitic chemical composition and 13% "fibers," as defined by a 3:1 aspect ratio [Wylie et al. 1993].

Since 1990, another carcinogenicity study of nonasbestiform amphibole minerals has been published. An IP injection study in rats used six samples of tremolite, including three asbestiform samples that induced mesothelioma in 100%, 97%, and 97% of challenged animals [Davis et al. 1991]. Two nonasbestiform tremolite samples resulted in mesotheliomas in 12% and 5% of the animals, and the response rate of 12% was above the expected background rate. Another sample that was predominantly nonasbestiform but reported to contain a small amount of asbestiform tremolite resulted in mesothelioma in 67% of animals. Of note, the nonasbestiform material associated with the 12% mesothelioma incidence and this latter material contained an approximately equal number of EMPs longer than 8 µm and thinner than 0.5 µm.

Studies of *in vitro* assays of various biological responses, some published before and some after 1990, have also found relative toxicities of asbestiform and nonasbestiform minerals that generally parallel the differences observed in the *in vivo* IP injection studies of tumorigenicity [Wagner et al. 1982; Woodworth et al. 1983; Hansen and Mossman 1987; Marsh and Mossman 1988; Sesko and Mossman 1989; Janssen et al. 1994; Mossman and Sesko 1990] A recent review of the literature concluded that low-aspect-ratio cleavage fragments of amphiboles are less potent than asbestos fibers [Mossman 2008].

In summary, there is more literature now than in 1990 pertaining to differential animal carcinogenicity and toxicity of EMPs from nonasbestiform amphiboles (e.g., acicular crystals, prismatic crystals, and cleavage fragments). The number of studies is limited and each study has limitations, but they suggest that nonasbestiform amphiboles might pose different risks than asbestos. More detailed discussion of these studies, including discussion of important limitations of the studies, can be found in Section 2.8.4 of this document.

2.7.1.1.3 Analytical Limitations

The third element that served as a basis for NIOSH's 1990 recommendation that nonasbestiform analogs of the asbestos minerals be encompassed by the REL was the inability to accurately and consistently distinguish asbestos fibers and nonasbestiform EMPs in samples of airborne particulate. The 1990 NIOSH testimony argued that asbestiform and nonasbestiform minerals can occur in the same geological area and that mixed exposures to airborne asbestos fibers and EMPs from the nonasbestiform analog minerals can occur at mining operations. The potential for mixed exposures can also occur downstream if the mined commodity contains both asbestiform and nonasbestiform minerals.

The 1990 NIOSH testimony further pointed out the lack of routine analytical methods for air samples that can accurately and consistently determine whether an individual EMP that meets the dimensional criteria of a countable particle is an asbestos fiber or a nonasbestiform EMP (e.g., acicular crystals, prismatic crystals, cleavage fragments).

Two analytical components of the NIOSH REL for airborne asbestos fibers are applied to air samples: the microscopic methods and the counting rules. The microscopic methods include the following:

- *Phase contrast microscopy* (PCM)—Analytical Method 7400 "A rules"—Asbestos and Other Fibers by PCM [NIOSH 1994a] is used to count all particles that are longer than 5 µm and have a length-to-width ratio equal to or greater than 3:1.

- *Transmission electron microscopy* (TEM)—Analytical Method 7402—Asbestos by TEM [NIOSH 1994b] is used as a supplement to the PCM method when there is uncertainty about the identification of elongate particles (EPs) that are counted. When TEM analysis is used for particle identification, only those EPs that are identified as "asbestos" and meet the dimensional criteria used by PCM (>0.25 µm width and >5 µm length) are counted as asbestos fibers. PCM counts can be adjusted to yield corrected asbestos fiber counts by multiplying them by the proportion of fibers determined by TEM to be asbestos.

There are several limitations of the use of PCM and TEM for asbestos analysis. PCM is stated to be limited to observing EPs with widths >0.25 µm and is not equipped for particle identification. TEM, although capable of resolving EPs with widths as small as 0.001 µm, frequently cannot differentiate nonasbestiform from asbestiform EMPs when the elemental composition is the same or when present in a heterogeneous mix of unknown particles. A potential limitation of TEM is that partial lengths of long fibers may not be observed because they intersect grid bars or lie partially outside the small field of view. However, this is likely to affect only a small number of observed particles, because very few particles are greater than 15 μm length, as evidenced in air samples from textile processing [Dement et al. 2009] and mining and milling operations [Gibbs and Hwang 1980]. A more important limitation of TEM is that, because only a small portion of the filter sample is being analyzed, airborne fiber concentrations might not be determined with certainty. Another limitation of both methods is that high concentrations of background dust collected on samples may interfere with fiber counting by PCM and particle identification by TEM.

Thus, the current PCM and TEM methods used for routine exposure assessment lack the capability to accurately count, size, and identify all EMPs collected on airborne samples. Further discussion of the analytical limitations and possible improvements are discussed in Section 2.8.

2.7.2 Clarification of the Current NIOSH REL

As described in the preceding sections, uncertainty remains concerning the adverse health effects that may be caused by nonasbestiform EMPs encompassed by NIOSH since 1990 in the REL for asbestos. Also as described in a preceding section, current analytical methods still cannot reliably differentiate between asbestos fibers and other EMPs in mixed-dust environments. NIOSH recognizes that its descriptions of the REL since 1990 have created confusion and caused many to infer that the additional covered minerals were included by NIOSH in its definition of "asbestos." NIOSH wishes to make clear that such nonasbestiform minerals are not "asbestos" or "asbestos minerals." NIOSH also wishes to minimize any potential future confusion by no longer referring to particles from the nonasbestiform analogs of the asbestos minerals as "asbestos fibers." However, as the following clarified REL makes clear, particles that meet the specified dimensional criteria remain countable under the REL for the reasons stated above, even if they are derived from the nonasbestiform analogs of the asbestos minerals.

With use of terms defined in this *Roadmap*, the NIOSH REL is now clarified as follows:

NIOSH has determined that exposure to asbestos fibers causes cancer and asbestosis in humans and recommends that exposures be reduced to the lowest feasible concentration. NIOSH has designated asbestos to be a "Potential Occupational Carcinogen"‡‡ and recommends that exposures be reduced to the lowest feasible concentration. The NIOSH REL for airborne asbestos fibers and related elongate mineral particles (EMPs) is 0.1 countable EMP from one or more covered minerals per cubic centimeter, averaged over 100 minutes, where

- a *countable elongate mineral particle* (EMP) is any fiber or fragment of a mineral longer than 5 μm with a minimum aspect ratio of 3:1 when viewed microscopically with use of NIOSH Analytical Method 7400 ('A' rules) or its equivalent; and

---

‡‡NIOSH's use of the term "Potential Occupational Carcinogen" dates to the OSHA classification outlined in 29 CFR 1990.103, and, unlike other agencies, is the only classification for carcinogens that NIOSH uses. See Section 6.1 for the definition of "Potential Occupational Carcinogen." The National Toxicology Program [NTP 2005], of which NIOSH is a member, has determined that asbestos and all commercial forms of asbestos are known to be human carcinogens based on sufficient evidence of carcinogenicity in humans. The International Agency for Research on Cancer (IARC) concluded that there was sufficient evidence for the carcinogenicity of asbestos in humans [IARC 1987b].

- a *covered mineral* is any mineral having the crystal structure and elemental composition of one of the asbestos varieties (chrysotile, riebeckite asbestos [crocidolite], cummingtonite-grunerite asbestos [amosite], anthophyllite asbestos, tremolite asbestos, and actinolite asbestos) or one of their nonasbestiform analogs (the serpentine minerals antigorite and lizardite, and the amphibole minerals contained in the cummingtonite-grunerite mineral series, the tremolite-ferroactinolite mineral series, and the glaucophane-riebeckite mineral series).

This clarification of the NIOSH REL for airborne asbestos fibers and related EMPs results in *no change* in counts made, as defined by NIOSH Method 7400 ('A' rules). However, it clarifies definitionally that EMPs included in the count are not necessarily asbestos fibers.

The NIOSH REL comprises two components. One component states the agency's intent about what minerals should be covered by the REL; the other component describes the sampling and analytical methods to be used for collecting, characterizing, and quantifying exposure to airborne particles from the covered minerals. Each of these components of the NIOSH REL is discussed in detail in later subsections of this *Roadmap*.

The NIOSH REL remains subject to change based on research findings that shed light on the toxicity of nonasbestiform amphibole EMPs covered by the REL and on the toxicity of other EMPs outside the range of those minerals currently covered by the REL. In addition, because of changes by the IMA in 1978 [Meeker et al. 2003] in how minerals (e.g., amphiboles) are identified and classified (optical microscopy to chemistry-based), a more extensive clarification of specific minerals covered by the NIOSH REL may be warranted. That more extensive clarification of covered minerals is beyond the scope of this *Roadmap* but will be addressed through additional efforts by NIOSH to encompass contemporary mineralogical terminology within the REL.

2.8 EMPs Other than Cleavage Fragments

2.8.1 Chrysotile

Chrysotile fibers consist of aggregates of long, thin, flexible fibrils that resemble scrolls or cylinders, and the dimensions of individual chrysotile fibers depend on the extent to which the material has been manipulated. Multi-fibril chrysotile fibers commonly split along the fiber length and undergo partial dissolution within the lungs [NRC 1984]. Such longitudinal splitting in the lung represents one way that air sample counts may underestimate the cumulative dose of fibers in the lung.

Epidemiological studies of chrysotile in Quebec mines [McDonald and McDonald 1997] and South Carolina textile mills [Dement et al. 1994; Hein et al. 2007] have produced very different estimates of the risk of cancer associated with exposure to chrysotile fibers. With the differences in fiber concentrations accounted for, the exposure in the textile mills appears to be associated with much higher risk than exposure in the mines. Several explanations for the difference in lung cancer risks observed in these two different workplaces have been proposed. One suggested explanation is that the textile workers were exposed to mineral oil. However, this explanation does not satisfactorily explain the differences [Stayner et al. 1996]. Considering that the textile mill workers were exposed to fibers considerably longer and thinner than those found in mines [Peto et al. 1982; Dement and Wallingford 1990], a more likely explanation is that the difference in risk may be due, at least in part, to dimensional differences in the particles to which workers were exposed. It has also been proposed that exposures in the textile mills were almost exclusively to chrysotile asbestos, whereas exposures in the mines were to a mixture of chrysotile asbestos fibers and EMPs of related nonasbestiform minerals (i.e., antigorite and lizardite) [Wylie and Bailey 1992]. Stayner et al. [1997] also point out, in comparing a number of epidemiological studies, that the variation in relative risk for lung cancer is often greater within an industry (e.g., mining or textile) than between varieties of asbestos.

Some have argued that pure chrysotile may not be carcinogenic and that increased respiratory cancer among chrysotile workers can be explained by the presence of tremolite asbestos as a contaminant of chrysotile [McDonald and McDonald 1997]. This is referred to as the amphibole hypothesis. However, several studies of workers using chrysotile with very little contamination by tremolite have demonstrated strong relationships between exposure to chrysotile and lung cancer. A study of chrysotile asbestos workers in China [Yano et al. 2001] found an age- and smoking-adjusted relative risk of 8.1 for lung cancer among highly exposed workers, compared to workers with low exposure to asbestos. The identified contamination of the chrysotile by tremolite was less than 0.001%. In the South Carolina textile mill study, a strong relationship between lung cancer and chrysotile exposure has been demonstrated [Dement et al. 1994; Hein et al. 2007]. The plant also processed less than 2000 lbs of crocidolite yarn per year from the early 1950s until 1972. This small amount of crocidolite was processed at one location, wet or lubricated with oil or graphite, to unite woven tape and braided packing [McDonald et al. 1983]. Since crocidolite was never carded, spun, or twisted, the predominant exposure at the plant was to chrysotile asbestos. A recent reanalysis by transmission electron microscopy (TEM) of archived fiber samples collected from the study facility in 1965 and 1968 identified only two amphibole fibers among 18,840 fiber structures (0.01%) in archived airborne dust samples from that textile mill study; the remainder were identified as chrysotile [Stayner et al. 2007]. Additionally, in fiber burden studies of malignant mesothelioma in humans, chrysotile fibers were often present in mesothelioma tissue, even in the absence of detectable amphibole fibers [Suzuki and Yuen 2001], though this alone does not constitute conclusive evidence of a role for chrysotile in the etiology of mesothelioma.

In addition to analyses of air samples from epidemiologic studies, UICC chrysotile B was analyzed, with special attention given to identifying tremolite contamination. Analysis of more than 20,000 fibers identified only particles with the characteristics of chrysotile, not tremolite [Frank et al. 1998]. Additionally, Kohyama et al. [1990] reported no amphibole asbestos was found in UICC chrysotile B at the detection level of 0.1%. This material has been used in an inhalation experiment that produced lung cancers and mesotheliomas [Wagner et al. 1974] and another study reporting mesotheliomas after intrapleural inoculation [Wagner et al. 1973].

A possible difference in risk for carcinogenicity between chrysotile and amphibole asbestos exposures has been investigated in animal model studies. In a one-year rat inhalation study, chrysotile samples were extremely fibrogenic and carcinogenic. Pulmonary carcinomas developed in approximately 25% of animals, and advanced interstitial fibrosis developed in lung tissue in 10% of all older animals, whereas intrapleural injection studies produced mesotheliomas in over 90% of animals [Davis et al. 1986]. It was noted that very little chrysotile remained in the lungs of the animals that survived longest following dust inhalation. From this it was suggested that chrysotile is very potent in rodents but, except where exposure levels are very high and of long duration, may be less hazardous to man because chrysotile fibers are removed from lung tissue more rapidly than are amphibole fibers. In a further analysis of the data, using regenerated dusts, Berman et al. [1995] found that although chrysotile and amphiboles do not differ in potency for lung tumor induction, mineralogy is a determinant of the relative potency of inhaled dusts for mesothelioma induction, and chrysotile is less potent than amphiboles.

Hodgson and Darnton [2000] reviewed the literature and estimated that, at exposure levels seen in occupational cohorts, the exposure-specific risk of mesothelioma from the three principal commercial asbestos types is broadly in the ratio 1:100:500 for chrysotile, amosite, and crocidolite, respectively, and the risk differential for lung cancer between chrysotile fibers and the two varieties of amphibole asbestos fibers is between 1:10 and 1:50. In 2009, Loomis et al. [2009] reported the results of a mortality study of chrysotile textile workers which found increased risks of both mesothelioma and lung cancer. The increase in mesothelioma and pleural cancer in this cohort was consistent with the average of 0.3% mesothelioma deaths in the cohorts that previously had been reported by Stayner et al. [1996] based on 12 studies of workers exposed to chrysotile in mining and processing plants, but higher than the estimates reported by Hodgson and Darnton [2000]. Based on the findings of Loomis et al. [2009], Hodgson and Darnton [2010] reanalyzed their data. They found that cumulative risk of mesothelioma for chrysotile-exposed asbestos workers in processing plants was approximately an order of magnitude greater than the risk they had previously reported for mines and processing plants combined, and commented that this risk is still at least an order of magnitude lower than that associated with exposure to amphibole asbestos [Hodgson and Darnton 2010].

A recent analysis of potency factors of various minerals estimated that the relative potency of chrysotile for producing mesothelioma ranged between zero and 1/200th that of amphibole asbestos and that amphibole asbestos and chrysotile were not equally potent in producing lung cancer [Berman and Crump 2008b]. However, the Asbestos Committee of EPA's Science Advisory Board expressed substantial concern that the scientific basis for a similar modeling approach being considered by EPA for risk assessment was weak and inadequate, and specifically cited the lack of available data to estimate TEM-specific levels of exposure for epidemiological studies used in the analysis [EPA 2008a]. Subsequently, EPA chose not to pursue this approach [EPA 2008b].

2.8.2 Asbestiform Amphibole Minerals

Asbestiform amphibole fibers consist of aggregates of long, thin, flexible fibrils that separate along grain boundaries between the fibrils. Because the fibril diameter of crocidolite is less than that of anthophyllite asbestos but the flexibility is greater, there is an indication that flexibility is a function of fibril diameter. As with chrysotile, the dimensions of individual amphibole asbestos fibers depend on the extent to which the material has been manipulated. There are conflicting reports on the splitting of amphibole fibers *in vivo*. Following intrapleural inoculation in rats, ferroactinolite and amosite have been reported to split longitudinally in the lung, resulting in increased numbers of thinner fibers with increased aspect ratio [Coffin et al. 1982, 1992; Cook et al. 1982], and crocidolite has also been observed to split longitudinally [Coffin et al. 1992]. In other studies, crocidolite splitting was not observed, even after several years [Bellman et al. 1987]. Longitudinal splitting of fiber bundles after entering the lung represents one way that air sample counts may underestimate the cumulative dose of fibers in the lung.

There is little scientific debate that the asbestiform varieties of the five commercially important amphibole asbestos minerals are carcinogenic and should be covered by regulations to protect workers. However, concerns have been raised that the current OSHA and MSHA asbestos regulations and even the current NIOSH REL for airborne asbestos fibers, which explicitly cover the asbestiform varieties of the six commercially important asbestos minerals, should be expanded to provide worker protection from exposure to EMPs from asbestiform minerals.

This concern is exemplified by exposures to winchite and richterite fibers at a vermiculite mine near Libby, Montana, where exposures to the these fibers have resulted in high rates of lung fibrosis and cancer among exposed workers, similar to the occurrence of asbestos-related diseases among asbestos-exposed workers in other industries [Amandus and Wheeler 1987; Amandus et al. 1987a,b; McDonald et al. 2004; Sullivan 2007; Rohs et al. 2008]. Workers at the mine and residents of Libby were exposed to EMPs identified (on the basis of the 1997 IMA amphibole nomenclature) as the asbestiform amphiboles winchite and richterite, as well as tremolite asbestos [Meeker et al. 2003].

The recently updated NIOSH cohort study of Libby workers found elevated SMRs for asbestosis (SMR 165.8; 95% CI 103.9–251.1), lung cancer (SMR 1.7; 95% CI 1.4–2.1), cancer of the pleura (SMR 23.3; 95% CI 6.3–59.5), and mesothelioma [Sullivan 2007]. An exposure-response relationship with duration of employment and total fiber-years cumulative exposure was demonstrated for both asbestosis and lung cancer. Significant excess mortality from non-malignant respiratory disease was observed even among workers with cumulative exposure <4.5 fibers/cc-years (i.e., a worker's cumulative lifetime exposure, if exposed to asbestos fibers at the current OSHA standard of 0.1 f/cm^3 over a 45-year working life). IARC stated mineral substances (e.g., vermiculite) that contain asbestos should also be regarded as "carcinogenic to humans" (Group 1) [IARC 2009]. Vermiculite from the Libby mine was used to produce loose-fill attic insulation, which remains in millions of homes around the United States, and homeowners and/or construction renovation workers (e.g., plumbers, cable installers, electricians, telephone repair personnel, and insulators) are potentially exposed when this loose-fill attic insulation is disturbed.

Because winchite and richterite are not explicitly listed among the six commercial asbestos minerals, it is sometimes assumed that they are not included in the regulatory definition for asbestos. However, some of what is now referred to as asbestiform winchite and richterite in the 1997 IMA nomenclature would have been accurately referred to as tremolite asbestos according to the analytical criteria (optical methods of identification) specified in the 1978 IMA nomenclature [Meeker et al. 2003]. Furthermore, an even greater portion of this richterite and winchite would have been identified as tremolite asbestos according to the optical methods of identification used prior to 1978. In fact, over the years, amphibole minerals from the Libby mine that are now referred to as winchite and richterite were identified by mineralogists using then-current terminology as soda tremolite [Larsen 1942], soda-rich tremolite [Boettcher 1966], and tremolite asbestos and richterite asbestos [Langer et al. 1991; Nolan et al. 1991]; NIOSH identified exposures to workers as tremolite in reports of the Libby mine epidemiological studies in the 1980s [Amandus and Wheeler 1987; Amandus et al. 1987a,b].

Similar to the situation in Libby, a study of a cluster of malignant mesothelioma cases in eastern Sicily has implicated an etiological role for an asbestiform amphibole in the fluoro-edenite series, initially identified as a mineral in the tremolite-actinolite series [Comba et al. 2003].

In the face of past and future nomenclature changes in the mineralogical sciences, workers need to be protected against exposures to pathogenic asbestiform minerals. The health and regulatory communities will need to carefully define the minerals covered by their policies and monitor the nomenclature changes to minimize the impact of these changes on worker protections.

2.8.3 Other Minerals of Potential Concern

By analogy to asbestiform amphiboles, there is reason to be concerned about potential for health risks associated with inhalational exposure to EMPs not covered by asbestos policies.

Erionite is perhaps the most worrisome known example [HHS 2005b]. An epidemic of malignant mesothelioma affecting several villages in Central Turkey has been studied for several decades [Baris et al. 1981]. Homes and other buildings in those villages were traditionally constructed of blocks of local volcanic stone containing erionite, a fibrous zeolite mineral. A recently published prospective mortality study has documented that mesothelioma accounts for over 40% of deaths among those residing in the affected villages [Baris and Grandjean 2006]. This localized epidemic of malignant mesothelioma produced an opportunity for a pedigree study that found a strong genetic influence on erionite-caused mesothelioma [Dogan et al. 2006]. As with exposure to asbestos, there is evidence that exposure to erionite causes other malignant tumors [Baris et al. 1996] and pleural plaques [Karakoca et al. 1997] in addition to mesothelioma. Likewise, as with amphiboles, the mineralogy of zeolites, including erionite, appears to be complicated and subject to misclassification [Dogan and Dogan 2008]. Although no clear epidemic of erionite-caused disease has been documented elsewhere, the mineral occurs in the intermountain west of the United States and a recent publication purports to be the first to report a case of erionite-associated malignant mesothelioma in North America [Kliment et al. 2009].

The International Agency for Research on Cancer (IARC) has considered evidence relevant to carcinogenicity for several EMPs [IARC 1977, 1987a, 1997]. In addition to the traditional commercial asbestos, IARC has made assessments that the evidence is sufficient to determine that both erionite and talc containing asbestiform fibers are human carcinogens (i.e., Group 1) [IARC 2009]; the evidence for carcinogenicity of nonasbestiform talc was judged to be insufficient to determine human carcinogenicity (i.e., Group 3) [IARC 1987a]. A Group 3 determination means that "the available studies are of insufficient quality, consistency or statistical power to permit a conclusion regarding the presence or absence of a causal association, or no data on cancer in humans are available" [IARC 1997]. On the basis of studies in rats, palygorskite (attapulgite) fibers longer than 5 mm were determined to be possibly carcinogenic to humans (Group 2B) [IARC 1997]. In experimental animals the evidence was limited for the carcinogenicity of long sepiolite fibers (>5 µm) and inadequate to assess carcinogenicity of nonerionite fibrous zeolites (including clinoptilolite, mordenite, and phillipsite) and wollastonite (Group 3) [IARC 1997]. These Group 3 determinations highlight the need for additional research on nonasbestiform EMPs.

2.9 Determinants of Particle Toxicity and Health Effects

Current recommendations for assessing occupational and environmental exposures to asbestos fibers rely primarily on dimensional and mineralogical characteristics. Dimension, which influences the deposition of EMPs in the lung, lung clearance mechanisms, and retention time in the lung, is an important determinant of toxicity. However, other particle characteristics, such as durability in lung fluids, chemical composition, and surface activity, may also play important roles in causing respiratory diseases. Research to elucidate what roles these EMP characteristics play in causing biological responses may help to provide better evidence-based recommendations for asbestos fibers and other EMPs.

2.9.1 Deposition

Deposition of airborne particles in the respiratory system is defined as the loss of particles from the inspired air during respiration. Clearance pertains to the removal of deposited particles by diverse processes over time, whereas retention is the temporal persistence of particles within the respiratory system [Morrow 1985]. The deposition of inhaled particles in the respiratory tract is a function of their physical characteristics (dimension and density), the anatomical and physiological parameters of the airways, and the rate and depth of respiration [Yu et al. 1986]. Although particle chemical composition does not play a role in deposition, respiratory clearance of all particle types is dependent on both physical and chemical characteristics of the particle. In addition, surface charge and hydrophilicity, as well as adsorbed materials (e.g., coatings on synthetic fibers) and other physical and chemical factors, determine whether small particles and fibers will agglomerate into larger, nonrespirable masses [ILSI 2005].

Depending on their physical characteristics, inhaled particles are differentially deposited in one of the following three respiratory system compartments: the extrathoracic region, consisting of the anterior and posterior nose, mouth, pharynx, and larynx; the bronchial region, consisting of the trachea, bronchi, and bronchioles down to and including the terminal bronchioles; and the alveolar-interstitial region, consisting of the respiratory bronchioles, alveolar ducts, and alveolar sacs.

Important parameters for the deposition of airborne particles are their aerodynamic and thermodynamic properties. Below a particle size of 0.5 µm aerodynamic equivalent diameter (AED), thermodynamic properties prevail. The AED of EPs is mostly determined by their geometric diameter and density. Deposition of EPs in an airway is strongly related to the orientation of the particles with respect to the direction of the air flow and is affected by the interrelationship of four major deposition mechanisms: impaction, interception, sedimentation, and diffusion [Asgharian and Yu 1988]. In a study to assess EP deposition in the tracheobronchial region, Zhou et al. [2007] evaluated the deposition efficiencies of carbon fibers (3.66 µm diameter), using two human airway replicas that consisted of the oral cavity, pharynx, larynx, trachea, and three to four generations of bronchi. Carbon fiber deposition was found to increase with the Stokes number, indicating that inertial impaction is the dominant mechanism. Also, fiber deposition in the tracheobronchial region was lower than that of spherical particles at a given Stokes number, indicating a greater likelihood for small-width EPs to move past the upper respiratory tract and reach the lower airways where diffusional deposition predominates [Yu et al.1986]. These results were confirmed by results of later studies evaluating the deposition of asbestos, using a similar tracheobronchial cast model [Sussman et al. 1991a, b]. The probability of deposition of a particle in a specific location in the airways is not the same as the probability of penetration to that region, and for particles in a certain range of AEDs, the difference between penetration and deposition may be substantial [ICRP 1994].

2.9.2 Clearance and Retention

A variety of mechanisms are associated with the removal of deposited particles from the respiratory tract [Warheit 1989]. Physical clearance of insoluble particles deposited in the lung is an important physiological defense mechanism that usually serves to moderate any risk that might otherwise be associated with exposure to particles. Inhaled particles that deposit on respiratory tract surfaces may be physically cleared by the tracheobronchial mucociliary escalator or nasal mucus flow to the throat, and then they may be either expectorated or swallowed.

Clearance depends upon the physicochemical properties of the inhaled particles, the sites of deposition, and respiratory anatomy and physiology. For example, inhaled insoluble particles with larger AEDs tend to deposit on the nasopharyngeal mucus and are generally cleared by sneezing or nose blowing or by flow into the oropharynx, where they are swallowed. Insoluble particles with smaller AEDs tend to deposit lower in the respiratory tract, with associated longer retention times. Those deposited in the alveolar region are subject to longer retention times than those deposited on the bronchial region [Lippmann and Esch 1988].

The most important process for removal of insoluble particles from the airways is mucociliary clearance, which involves a layer of mucus propelled by the action of ciliated airway cells that line the trachea, bronchi, and terminal bronchioles [Warheit 1989]. The mucociliary transport system is sensitive to a variety of agents, including cigarette smoke and ozone [Vastag et al. 1985]. These toxicants affect the speed of mucus flow and consequent particle clearance by altering ciliary action and/or modifying the properties and/or amount of mucus. Chronic exposure to cigarette smoke has been shown to cause a prolonged impairment of particulate clearance from the bronchial region. This impaired clearance is associated with increased retention of asbestos fibers in the bronchi, where they stimulate inflammatory processes in the bronchial epithelium [Churg and Stevens 1995; Churg et al. 1992].

Because the alveolar region of the lung does not possess mucociliary clearance capability, particles (generally <2 µm AED) deposited in this region are cleared at a much slower rate than particles deposited in the bronchial region. Particles that are soluble may dissolve and be absorbed into the pulmonary capillaries, whereas insoluble particles may physically translocate from the alveolar airspace [Lippmann et al. 1980; Lippmann and Schlesinger 1984; Schlesinger 1985]. Most insoluble EPs that deposit in the alveolar regions are phagocytized (i.e., engulfed) by alveolar macrophages. Macrophages contain lysosomes packed with digestive enzymes, such as acid hydrolases, at acidic pH levels. Lysosomal contents are capable of digesting many—though not all—types of phagocytized particles. Alveolar macrophages that have phagocytized particles tend to migrate to the bronchoalveolar junctions, where they enter onto the mucociliary escalator for subsequent removal from the lung [Green 1973]. It has been postulated by some investigators that dissolution of particles within macrophages is a more important determinant of long-term clearance kinetics for many mineral dusts than is mucociliary transport and the migratory potential of lung macrophages [Brain et al. 1994]. However, certain circumstances can disrupt the normal phagosomal function of alveolar macrophages. One such circumstance involves the toxic death of macrophages initiated by highly reactive particle surfaces (e.g., crystalline silica particles). Another such circumstance involves overwhelming the capacity of macrophages by an extreme burden of deposited particles, sometimes referred to as "overload," even by particles that would be considered "inert" at lower doses. A third type of circumstance, typified by asbestos fibers, involves EPs that, even though having a small enough AED (defined primarily by particle width) to permit deposition in the alveolar region, cannot be readily phagocytized because particle length exceeds macrophage capacity. When alveolar macrophages attempt to phagocytize such EPs, they cannot completely engulf them (sometimes referred to as "frustrated phagocytosis"), and lysosomal contents are released into the alveolar space. Frustrated phagocytosis can initiate a process in which reactive oxygen species (ROS) are generated, stimulating the induction of tumor necrosis factor-alpha (TNF-α). TNF-α is considered an inflammatory and fibrogenic cytokine that plays an important role in the pathogenesis of pulmonary fibrosis [Blake et al. 1998].

All three types of disruption of normal macrophage function contribute to decreased particle clearance rates and can result in inflammation of the alveolar spaces. In addition, particles that are not phagocytized in the alveoli can translocate to the lung interstitium, where they may be phagocytized by interstitial macrophages or transported through the lymphatic system to pulmonary lymph nodes [Lippmann et al. 1980; Lippmann and Schlesinger 1984; Schlesinger 1985; Oberdorster et al. 1988]. Tran and Buchanan [2000] have reported findings suggesting that sequestration of particles in the interstitial compartment is more prominent in exposed humans than is the observed retention of particles due to overload in animal studies. The importance of interstitialization in humans is consistent with the kinetic differences observed in lung clearance rates in humans and rats. The first-order rate coefficient for alveolar clearance is approximately 1 order of magnitude faster in rats than in humans [Snipes 1996], which may allow for greater interstitialization of particles in humans at all levels of lung dust burden. These findings indicate that adjustment of kinetic differences in particle clearance and retention is required when using rodent data to predict lung disease risks in humans and that current human lung models underestimate working lifetime lung dust burdens in certain occupational populations [Kuempel et al. 2001].

Evidence from *in vivo* studies in rodents and *in vitro* studies indicates that EPs (vitreous glass and EMPs) with a length equal to or greater than the diameter of rodent lung macrophages (about 15 µm) are most closely linked to biological effects observed in rodent lungs [Blake et al. 1998]. Alveolar macrophages appear to be capable of phagocytizing and removing EPs shorter than approximately 15 µm, either by transport to the mucociliary system or to local lymph channels. With increasing length above approximately 15 µm, alveolar macrophages appear to be increasingly ineffective at physical removal, resulting in differential removal rates for EPs of different lengths. Although EP lengths greater than 15 µm appear to be associated with toxicity in experimental studies with rodents, a "critical" length for toxicity in humans is probably greater than 15 µm [Zeidler-Erdely et al. 2006]. For long EPs that cannot be easily cleared by macrophages, biopersistence in the lung is influenced by the ease with which the EPs break into shorter lengths.

2.9.3 Biopersistence and Other Potentially Important Particle Characteristics

It has been hypothesized that the differences in crystalline structure between amphibole asbestos fibers and amphibole cleavage fragments account for apparent differences in toxicological response to these particles. Cleavage fragments that meet the dimensional criteria for countable particles under federal regulatory policies for asbestos fibers are generally shorter and wider than asbestos fibers [Siegrist and Wylie 1980; Wylie 1988]. This dimensional difference between populations of asbestos fibers and populations of cleavage fragments might contribute to generally shorter biopersistence in the lung for cleavage fragments than for asbestos fibers. Asbestos fibers also tend to separate longitudinally once deposited in the lung, thus increasing the total number of retained fibers without an accompanying reduction in lengths of the retained fibers [NRC 1984]. In contrast, cleavage fragments tend to break transversely because of dissolution of their weaker crystalline structure, resulting in shorter particles that can be more easily cleared through phagocytosis and mucociliary clearance [Zoltai 1981]. The impact of these structural differences on solubility in lung fluids warrants study, because substantial differences in solubility in lung fluids between asbestos fibers and other EMPs (including amphibole cleavage fragments) could translate into differences in toxicity.

2.9.3.1 Biopersistence

Dissolution of EPs in the lung is a poorly understood process that is dependent on particle characteristics, biological processes, and concomitant exposure to other particulates. The ability of an EP to be retained and remain intact in the lung is considered an important factor in the process of an adverse biological response. EPs of sufficient length that remain intact and are retained in the lung are thought to pose the greatest risk for respiratory disease. The ability of an EP to reside long-term in the lung is generally referred to as biopersistence. Biopersistence of EPs in the lung is a function of site and rate of deposition, rates of clearance by alveolar macrophages and mucociliary transport, solubility in lung fluids, breakage rate and breakage pattern (longitudinal or transverse), and rates of translocation across biological membranes. The rates of some of these processes can affect the rates of other processes. For example, a high rate of deposition in the alveolar region could potentially overwhelm macrophage clearance mechanisms and increase the rate of translocation to the lung interstitium.

The persistence of an EP in the lung is influenced by changes that may occur in its dimension, surface area, chemical composition, and surface chemistry. Differences in any of these characteristics can potentially result in differences in clearance and retention and affect toxic potential. For example, EPs too long to be effectively phagocytized by alveolar macrophages will tend to remain in the alveolar compartment and be subjected to other clearance mechanisms, including dissolution, breakage, and translocation to interstitial sites and subsequently to pleural and other sites.

The durability of EPs residing in the lung is an important characteristic influencing biopersistence. An EP's durability is generally measured by its ability to resist dissolution and mechanical disintegration after being subjected to lung extracellular fluid (approximate pH of 7) and lysosomal fluids (approximate pH of 5). EPs that are more soluble will be less biopersistent, and thicker EPs may take longer to dissolve than thinner EPs, all else being equal. For example, long, thin EPs that are not very durable could dissolve and/or fragment into shorter EPs, increasing their probability of being cleared from the lung and thus potentially decreasing lung retention time and risk for fibrotic or neoplastic effects. Some EPs, such as certain types of glass fibers, are fairly soluble in lung fluid and are cleared from the lung in a matter of days or months. Other EPs, such as amphibole asbestos, can remain in the lung for decades. It has been suggested that some types of EPs may alter the mobility of macrophages and the translocation of EPs to the pleura or lymph nodes [Davis 1994]. No relationship has been established between biopersistence of EPs in the lung and the risk of induction of genetic and epigenetic changes that may lead to cancer [Barrett 1994]. Although some evidence indicates that durability may be a determinant of toxicity for various synthetic vitreous fibers (SVFs), EMPs need to be evaluated to determine whether they conform to this paradigm [ILSI 2005].

Measurement of the biopersistence of various EMPs has been suggested as a means for estimating their relative potential hazard. Short-term inhalation and intratracheal instillation studies have been used to determine the biopersistence of various SVFs and asbestos fibers. Animal inhalation studies are preferred over animal tracheal instillation studies to assess biopersistence because they more closely mimic typical human exposure. The European Commission has adopted specific testing criteria that permit the results from either short-term biopersistence studies or chronic animal studies to be used as a basis for determining carcinogenicity [European Commission 1997].

Several animal inhalation studies have indicated that oncogenic potential of long SVFs can be determined by their biopersistence [Mast et al. 2000; Bernstein et al. 2001; Moolgavkar et al. 2001]. It has been suggested that a certain minimum persistence of long EPs is necessary before even minute changes appear in the lungs of exposed animals [Bernstein et al. 2001]. Furthermore, Moolgavkar et al. [2001] have suggested that EP-induced cancer risk, in addition to being a linear function of exposure concentration, is also a linear function of the weighted half-life of EPs observed in inhalation studies with rats. Also, dosimetry models for rodents and humans indicate that, on a normalized basis, EP clearance rates are lower in humans than in rats [Maxim and McConnell 2001] and that EPs frequently sequester in the interstitial compartment of humans [Snipes

1996; Tran and Buchanan 2000]. Thus, results from chronic inhalation studies with rodents exposed to EPs may underestimate risks for humans, and adjustment for kinetic differences in particle clearance and retention in rats is required to predict lung disease risks in humans [Kuempel et al. 2001].

Studies using *in vitro* assays have been conducted with various SVFs and silicate minerals to determine the dissolution rate in simulated lung and lysosomal fluids [Hume and Rimstidt 1992; Werner et al. 1995; Hesterberg and Hart 2000; Jurinski and Rimstidt 2001]. In vitro dissolution studies can provide a rapid and more controlled alternative to classic long-term toxicity testing in animals and could provide useful information when performed as companion experiments with *in vivo* studies if conditions of exposure and test agent can be made similar. The design of *in vitro* assays is intended to mimic the biological conditions that exist in the lung once the EP comes into contact with lung tissue or macrophages. Although uncertainties exist about the specific physiological processes that occur in the lung, results from *in vitro* assays can provide some insight into the chemical reactions that influence EP dissolution. For example, it appears that EP (e.g., glass fiber) dissolution occurs more readily when the EP is in contact with a fluid that is undersaturated with respect to the EP's composition. The condition of undersaturation must be maintained at the EP's surface for dissolution to continue. If an EP is surrounded by a saturated or supersaturated solution (compared to the EP composition), then no further dissolution occurs.

The results from many *in vitro* experiments demonstrate different patterns of dissolution for most of the tested EP types (i.e., glass, asbestos) under various test conditions. This effect was most notable in those experiments where different pH conditions were used. Fluid pH appears to influence the creation of complexes from the leached elements of the EP, which in turn alters the rate of solubility. Chrysotile fibers tend to dissolve readily in acids because of the preferential leaching of magnesium (Mg) from the fiber. The leaching of Mg from tremolite and anthophyllite and of sodium from crocidolite also occurs more readily in acid conditions.

The rate of EP dissolution has also been observed to be affected by differing internal and surface structures. EPs with porous or rough surfaces have larger surface areas than smooth EPs with the same gross dimensions. These larger surface areas interact more readily with the surrounding medium because of the greater number of sites where solute molecules can be absorbed. EMPs with cleavage plane surfaces will contain varying degrees of defects; the higher the number of surface defects, the greater the potential instability of the particle. Dissolution of these types of EMPs is typically initiated where surface vacancies or impurities are present [Searl 1994]. Chrysotile asbestos is an example of a sheet silicate made up of numerous fibrils composed of tightly bound rolled layers of Mg hydroxide. These Mg hydroxide layers are readily leached by acid solutions within human tissues [Spurny 1983], causing disintegration of the fibril's crystalline structure. In contrast, the amphibole asbestos minerals are chain silicates with a crystalline structure comprising alkali and alkali earth metals that are tightly bound, making the fibers less susceptible to dissolution. In contrast to the crystalline structure of the asbestos fibers, some high-temperature glass fibers are more stable than chrysotile fibers because they are composed of silicate chains, sheets, and frameworks [Searl 1994]. The absence of cleavage planes or structural defects in glass fibers limits the degree to which fluids can penetrate their interior to promote dissolution. Chrysotile fibers were found to be less durable in rat lungs than some high-temperature SVFs [Bellmann et al. 1987; Muhle et al. 1987] but more durable in physiological solutions than some refractory ceramic fibers (RCFs) [Scholze and Conradt 1987].

EP surface characteristics (e.g., structural defects, porosity) and composition affect not only the rate of dissolution but also the manner in which dissolution occurs. In some instances, surface dissolution will cause alterations in internal structure sufficient to cause mechanical breakage. In some studies, slagwools and rockwools exposed to water developed irregular surfaces, creating stress fractures that caused transverse breakage [Bellmann et al. 1987]. Similar occurrences of glass fiber breakage have been observed when there was leaching of alkaline elements [Searl 1994].

Results from *in vitro* and short-term *in vivo* studies conducted with various EMPs and SVFs provide some confirmation that persistence of EPs in the lung is influenced by particle durability [Bernstein et al. 1996]. However, other evidence suggests that, because of the relatively short biodurability of chrysotile fibers, any damage to the lung tissue caused by chrysotile fibers must be initiated soon after exposure [Hume and Rimstidt 1992], suggesting that biopersistence of EPs in the lung may be only one of many factors that contribute to biological response. A better understanding of the factors that determine the biological fate of EMPs deposited in the lung is critical to understanding the mechanisms underlying differences in toxic potential of various EMPs of different dimensions and compositions. Because biopersistence of EMPs is thought to play an important role in the development of disease, it may eventually prove to be an important characteristic to incorporate into occupational safety and health policies concerning exposures to EMPs.

2.9.3.2 Other Potentially Important Particle Characteristics

Surface composition and surface-associated activities have been suggested as factors affecting the potential for disease induction by EMPs (e.g., asbestos) [Bonneau et al. 1986; Kane 1991; Jaurand 1991; Fubini 1993]. For nonelongate respirable mineral particles (e.g., crystalline silica), surface composition and surface interactions can directly and profoundly affect *in vitro* toxicities and *in vivo* pathogenicity; they can also directly cause membranolytic, cytotoxic, mutagenic, or clastogenic damage to cells and have been shown to induce fibrogenic activities in animals and humans. Investigation is warranted to confirm that these effects of surface composition and surface interactions also apply to EMPs. One strategy is to determine the effects of well-characterized surface modification of different types of EMPs on cell-free interactions with biological materials, *in vitro* cellular cytotoxicities or genotoxicities, and pathology in animal models.

Surface properties of mineral fibers and other EMPs may have direct impact on cytotoxic or genotoxic mechanisms responsible for fibrogenic or carcinogenic activity. Chemical surface modification of asbestos fibers has been shown to affect their cytotoxicity [Light and Wei 1977a,b; Jaurand et al. 1983; Vallyathan et al. 1985]. Although asbestos fibers clearly can be carcinogenic, they are not consistently positive in genotoxicity assays; their principal damage is chromosomal rather than gene mutation or DNA damage [Jaurand 1991]. One study linked cytotoxicity with *in vitro* mammalian cell transformation [Hesterberg and Barrett 1984]; thus, surface factors affecting cytotoxicity might affect potential for inducing some genotoxic activities. However, modifying the surface of a well-characterized sample of chrysotile fibers by depleting surface Mg while retaining fiber length did not result in a significant quantitative difference for *in vitro* micronucleus induction between the native and surface-modified materials, both of which were positive in the assay [Keane et al. 1999].

The surface of mineral fibers and other EMPs also might be an indirect but critical factor in the manifestation of pathogenic activity. EMP surfaces may be principal determinants of EMP durability under conditions of *in vivo* dissolution in biological fluids. As such, they would be a controlling factor in biopersistence, critical to the suggested mechanisms of continuing irritation or inflammatory response in causing fibrosis or neoplastic transformation.

2.9.4 Animal and *In Vitro* Toxicity Studies

Over the last half-century, *in vivo* animal model studies have explored induction of lung cancer, mesothelioma, and pulmonary fibrosis by asbestos fibers and other EMPs following intrapleural, intraperitoneal, or inhalation challenge. Numerous cell-free, *in vitro* cellular, and *in vivo* short-term animal model studies have been pursued, attempting to (1) examine tissue and cellular responses to EMPs and impact of EMP conditioning on these responses; (2) identify and evaluate interactions and mechanisms involved in pathogenesis; and (3) seek morphological or physicochemical EMP properties controlling those mechanisms. These short-term studies provide an evolving basis for design or interpretation of higher-tier chronic exposure studies of selected EMPs.

Some of the short-term studies have addressed the following:

- the general question of extrapolating human health effects from *in vivo* animal model studies;
- the physiological relevance of *in vitro* cellular studies of EMP toxicities;
- the association of EMP dimensions with pathology demonstrated in animal model studies;
- the potential mechanisms and associated EMP properties responsible for initiating cell damage;
- the extensive information now available on a "central dogma" of subsequent intracellular biochemical pathway stimulation leading to toxicity or intercellular signaling in disease promotion; and
- the use of these mechanistic paradigms to explain specific questions of
  — differences between the activities of asbestiform and nonasbestiform EMPs, including seemingly anomalous differences between some *in vitro* and *in vivo* EMP activities;
  — differences between the activities of erionite fibers and amphibole asbestos fibers; and
  — the possibility of EMP-viral co-carcinogenesis.

Several reviews and recommendations for animal model and cellular studies on these issues have been developed by expert workshops and committees. Early studies on the carcinogenicity of asbestos and erionite fibers were reviewed by IARC [1977, 1987a,b], and SVFs were reviewed more recently [IARC 2002]. Short-term *in vivo* and *in vitro* studies to elucidate mechanisms of fiber-induced genotoxicity and genetic mechanisms affecting fiber-induced lung fibrosis have been extensively reviewed.

A review for the EPA by an international working group assembled in 2003 provides an update on short-term assay systems for fiber toxicity and carcinogenic potential [ILSI 2005], and two additional reviews discuss the fiber genotoxicity literature up to the current decade [Jaurand 1997; Schins 2002].

2.9.4.1 Model Systems Used to Study EMP Toxicity

The paucity of information on human health effects of some new synthetic EPs has led to renewed considerations of the value and limitations of animal model studies, as well as the interpretability of intrapleural, intraperitoneal, or inhalation challenge methods of animal model tests to make predictions of human health effects [IARC 2002]. One analysis concluded that rat inhalation is not sufficiently sensitive for prediction of human carcinogenicity by EMPs other than asbestos fibers [Muhle and Pott 2000]. Another review concluded that there are significant interspecies differences between the mouse, hamster, rat, and human, with the available evidence suggesting that the rat is preferable as a model for the human, noting that rats develop fibrosis at comparable lung burdens, in fibers per gram of dry lung, to those that are associated with fibrosis in humans. The review suggested that, on a weight-of-evidence basis, there is no reason to conclude that humans are more sensitive to fibers than rats with respect to the development of lung cancer [Maxim and McConnell 2001]. However, others suggest that, because inhaled particles frequently sequester in the interstitial compartment of humans, alveolar clearance is approximately one order of magnitude slower in humans than in rats [Snipes 1996; Tran and Buchanan 2000]. Those comparisons imply that results of inhalation studies with rats exposed to particles underestimate the risk for humans and that adjustment for kinetic differences in particle clearance and retention in rats is required to predict lung disease risks in humans [Kuempel et al. 2001].

How the results of *in vitro* tests that use cells or organ cultures apply to humans has been questioned because of differences in cell types and species-specific responses. It is difficult to isolate and maintain epithelial or mesothelial cells for use as models. Interpretation of *in vitro* test results may be limited because *in vitro* models may not consider all processes, such as clearance or surface conditioning, which occur *in vivo*. A major deficiency of *in vitro* systems is that biopersistence is not easily addressed. In addition to the usual exposure metric of mass, experimental designs should also include exposure metrics of EMP number and surface area [Mossman 2008; Wylie et al. 1997].

As frequently performed, *in vitro* assays of mineral particle-induced damage, measured by cell death or cytosolic or lysosomal enzyme release, do not adequately model or predict the results of *in vivo* challenge or epidemiological findings. For example, respirable aluminosilicate clay dust is as cytotoxic as quartz dust in such *in vitro* assays, whereas quartz, but not clay, is strongly fibrogenic *in vivo* [Vallyathan et al. 1988].

2.9.4.2 Studies on Effects of Fiber Dimension

Early animal inhalation studies showed that chrysotile fibers induced fibrosis, hyperplasia of lung epithelial cells, and carcinomas in mice [Nordman and Sorge 1941] and tumors in rats [Gross et al. 1967]. Another study revealed lung carcinomas and mesotheliomas in rats after inhalation exposure to asbestos fiber samples of amosite, anthophyllite, crocidolite, and chrysotile [Wagner et al. 1974]. The effects of fiber length, width, and aspect ratio on carcinogenicity were addressed in a seminal study using a pleural surface implantation method of challenge in the rat [Stanton et al. 1977, 1981]. Tests were performed on 72 durable EPs: 13 crocidolites; 22 glasses; 8 aluminum oxide sapphire whiskers; 7 talcs; 7 dawsonites; 4 wollastonites; 2 asbestos tremolites; 1 amosite; 2 attapulgites; 2 halloysites; 1 silicon carbide whisker; and 3 titanates. The incidence of malignant mesenchymal neoplasms a year after implantation correlated best with EPs that were longer than 8 μm and no wider than 0.25 μm, with relatively high correlations with EPs longer than 4 μm and no wider than 1.5 μm. This suggested that carcinogenicity of durable EPs depends on dimension and durability, rather than physicochemical properties. This is sometimes referred to as the Stanton hypothesis and has been the subject of continuing research. Reanalysis of the dimensions of seven of the crocidolite samples used in the 1981 study found that tumor probability was significantly correlated with the number of index particles (defined as particles longer than 8 μm and no wider than 0.25 μm), but the coefficient was low enough to suggest that factors other than size and shape play a role in carcinogenic effects of durable EPs [Wylie et al. 1987]. Further analysis confirmed the number of such index particles as the primary dimensional predictor of tumor incidence, but the correlation was increased when the data were analyzed by separate mineral types [Oehlert 1991]. These analyses suggested that mineral type is important, which is counter to the Stanton hypothesis.

Lippmann [1988] reviewed data from animal models exposed by instillation or inhalation of EMPs of defined size distributions, along with data on human lung fiber burden and associated effects, and concluded that (1) asbestosis is most closely associated with the surface area of retained EMPs; (2) mesothelioma is most closely associated with numbers of EMPs longer than about 5 μm and thinner than about 0.1 μm; and (3) lung cancer is most closely associated with EMPs longer than about 10 μm and thicker than about 0.15 μm. Lippmann [1988] also stated that the risk of lung cancer is associated with a substantial number of fibers longer than 10 μm with widths between 0.3 μm and 0.8 μm. A more recent review of the response to asbestos fibers of various lengths in animal models, along with data from studies of human materials, concluded that asbestos fibers of all lengths induce pathological responses; it suggested caution when attempting to exclude any subpopulation of inhaled asbestos fibers, on the basis of length, from being considered contributors to the development of asbestos-related diseases [Dodson et al. 2003].

2.9.4.3 Initiation of Toxic Interactions

A first question in seeking a full understanding of EMP properties and mechanisms responsible for fibrosis, lung cancer, or mesothelioma risks is the identity of initiating toxic interactions and the morphological, physical, and/or chemical properties of EMPs controlling them. Among proposed initiating mechanisms are these: (1) EMP surfaces generate ROS (even *in vitro* in the absence of cells), which are the primary toxicants to cells; (2) EMP surfaces are directly membranolytic or otherwise directly cytotoxic or genotoxic to components of the cell, as are some nonelongate mineral particles, and that damage can cause necrosis, apoptosis, mutation, or transformation directly or by responsive cellular production of secondary reactive intermediates; and (3) EMP morphology itself can result in frustrated phagocytosis with an anomalous stimulation or release of ROS or other toxic reactive species.

2.9.4.3.1 Reactive Oxygen Species

Asbestos fibers can generate ROS or reactive nitrogen species in *in vitro* systems through direct aqueous-phase surface chemical reactions, as well as by stimulating secondary release of reactive species from cells. Electron spin resonance with use of spin-trapping techniques showed that crocidolite, chrysotile, and amosite asbestos fibers were all able to catalyze the generation of toxic hydroxyl radicals in a cell-free system containing hydrogen peroxide, a normal byproduct of tissue metabolism, and that the iron chelator desferroxamine inhibited the reaction, indicating a major role for iron in the catalytic process [Weitzman and Graceffa 1984]. ROS generated by some EMP surfaces in cell-free media may provide toxicants to initiate cell structural or functional damage, including chromosomal or DNA genetic damage or aneuploidy from spindle apparatus damage. They also may activate cellular signaling pathways that promote cell proliferation or transformation. Research has investigated the possible roles of iron in this reactivity and the roles of released versus surface-borne iron.

Asbestos fibers can cause lipid peroxidation in mammalian cells and artificial membranes that can be prevented by removal of catalytic iron. Reduction of crocidolite cytotoxicity by certain antioxidants (including superoxide dismutase [SOD], a depletor of superoxide anion [SO]; catalase, a scavenger of hydrogen peroxide [H_2O_2]; dimethylthiourea [DMTU], a scavenger of the hydroxyl radical [•OH]; and desferroxamine, an iron chelator) suggested that iron is involved in the generation of ROS through a modified Haber-Weiss Fenton-type reaction resulting in the production of hydroxyl radical (e.g., from SO and H_2O_2 generated during phagocytosis) [Goodglick and Kane 1986; Shatos et al. 1987]. Such scavenging or chelation prevented DNA strand breakage in cells *in vitro* by crocidolite fibers [Mossman and Marsh 1989].

In a cell-free study of five natural and two synthetic fibers, erionite, JM code 100 glass fibers, and glass wool were the most effective initiators of hydroxyl radical formation, followed by crocidolite, amosite, and chrysotile fibers. Hydroxyl radical formation activity showed positive correlations with tumor rates in rats challenged by intrapleural injection and with human mesothelioma mortality rates, but not with tumor rates in rats challenged by intraperitoneal injection [Maples and Johnson 1992]. SO-produced ROS then might induce DNA oxidative damage, measured as elevated 8-hydroxydeoxyguanosine (8-OHdG). In cell-free systems, the crocidolite-induced increase of 8-OHdG in isolated DNA was enhanced by addition of H_2O_2 and diminished by addition of desferroxamine [Faux et al. 1994]. However, de-ironized crocidolite fibers incubated in a cell-free system induced twice the 8-OHdG oxidative damage to DNA as untreated crocidolite fibers. In parallel rat exposures, the combination of de-ironized crocidolite fibers plus Fe_2O_3 resulted in mesothelioma in all animals, whereas mesothelioma developed in only half the animals injected with crocidolite fibers alone and none of the animals injected with Fe_2O_3 alone [Adachi et al. 1994]. Other research suggested that unreleased fiber-surface-bound iron is important to the reactivity; long fibers of amosite and crocidolite both caused significant dose-dependent free radical damage to cell-free phage DNA, suppressible by the hydroxyl radical scavenger mannitol and by desferroxamine, but short RCFs and man-made vitreous fibers (MMVFs) did not, while releasing large quantities of Fe(III) iron [Gilmour et al. 1995]. Crocidolite fibers induced mutations in peritoneal tissue *in vivo* in rats, most prominently guanine-to-thymine

(G-to-T) trans-versions known to be induced by 8-OHdG; this was interpreted as strong evidence for the involvement of ROS or reactive nitrogen species in crocidolite-induced mutagenesis *in vivo*, consistent with *in vitro* and cell-free studies [Unfried et al. 2002]. In contrast to glass fiber, crocidolite fiber intratracheal instillation in rats increased 8-OHdG levels in DNA at one day and in its repair enzyme activity at seven days. This *in vivo* activity is consistent with asbestos- and MMVF-induced increases of 8-OHdG oxidative damage *in vitro* [Yamaguchi et al. 1999].

2.9.4.3.2 Membrane Interactions

Many mineral particles, elongate or not, can directly cause membranolysis or other cytotoxic responses without necessarily invoking extracellular generation of ROS. Mechanisms of cell damage by EMPs, independent of ROS formation, have been proposed to involve direct interactions of particle surface functional groups (e.g., silicon or aluminum or magnesium) with lipoproteins or glycoproteins of the cell membrane. It has been suggested that silica particle cytotoxicity to macrophages is due to distortion and disruption of secondary lysosomal membranes by phagocytosed particles whose surface silanol groups hydrogen-bond to membrane lipid phosphates, but that chrysotile-induced cellular release of hydrolytic enzymes is due to surface magnesium interacting ionically with sialic acid residues of membrane glycoproteins, inducing cation leakage and osmotic lysis [Allison and Ferluga 1977]. Chrysotile fibers cause lysis of red blood cells. EM indicates that cell membranes become wrapped around the fibers and that cell distortion and membrane deformation correlate with an increase in the intracellular ratio of sodium to potassium ions. Cell pretreatment with neuraminidase prevents fiber-cell binding, suggesting mediation by cell membrane glycoproteins [Brody and Hill 1983]. However, chrysotile and crocidolite fibers both induced increased membrane rigidity in model unilamellar vesicles made of saturated dipalmitoyl phosphatidylcholine (DPPC), suggesting that lipid peroxidation is not involved in membrane rigidity induced by asbestos [Gendek and Brody 1990]. Silicate slate dust and chrysotile fibers both induced hemolysis *in vitro* and peroxidation of polyunsaturated membrane lipids. However, poly(2-vinylpyridine N-oxide) (PVPNO) and DPPC surface prophylactic agents suppressed lysis but not peroxidation, whereas SOD and catalase did the reverse; and lysis was much faster than peroxidation. This suggested that membrane lysis and peroxidation are independent processes [Singh and Rahman 1987]. However, either mechanism may be involved in membrane damage by EMPs; and seemingly disparate findings suggest uncharacterized details of EMP properties or of cellular or mineral conditioning under test conditions may be important.

In *in vitro* studies, quartz dust and chrysotile fibers induced loss of viability and release of lactate dehydrogenase (LDH) from alveolar macrophages. DPPC reduced these activities of the quartz but not of the asbestos [Schimmelpfeng et al. 1992]. DPPC is adsorbed from aqueous dispersion in approximately equal amounts on a surface area basis, about 5 mg phospholipid per square meter, by asbestos fibers [Jaurand et al. 1980] and by nonfibrous silicate particles [Wallace et al. 1992]; this is close to the value predicted by mathematical modeling of an adsorbed bilayer [Nagle 1993]. In the case of silica or clay membranolytic dusts, this adsorption fully suppresses their activity until toxicity is manifested as the prophylactic surfactant is digested from the particle surface by lysosomal phospholipase enzyme; mineral-specific rates of the process suggest a basis for differing fibrogenic potentials of different types of mineral particles [Wallace et al. 1992].

Samples of intermediate-length and short-length NIEHS chrysotile were compared, with and without DPPC lung surfactant pretreatment, for micronucleus induction in Chinese hamster lung V79 cells *in vitro*. Increase in micronuclei frequency and multinuclear cell frequency were induced by all samples, with the greatest micronucleus induction by untreated intermediate-length chrysotile fibers and with greater activity for untreated versus treated short chrysotile fibers. Cell viability was greater for treated fibers [Lu et al. 1994]. NIEHS intermediate-length chrysotile was mildly acid-treated to deplete surface-borne magnesium while only slightly affecting fiber length. Challenge of Chinese hamster lung fibroblast cells *in vitro* for micronucleus induction found no significant difference between the treated and untreated samples, supporting a model of chemically nonspecific chromosomal and spindle damage effects [Keane et al. 1999]. Chrysotile fiber induction of mucin secretion in a tracheal cell culture was inhibited by using lectins to block specific carbohydrate residues on the cell surface; leached chrysotile was inactive, suggesting that the surface cationic magnesium of chrysotile was responsible for interaction with cell surface glycolipids and glycoproteins [Mossman et al. 1983]. However, complete removal of accessible sialic acid residues from erythrocytes did not inhibit hemolysis by chrysotile fibers, suggesting that chrysotile fibers can induce lysis by interaction with some other component of the cell [Pelé and Calvert 1983].

2.9.4.3.3 Morphology-mediated Effects

A third possible mechanism for damage by EMP principally involves morphology. The possibility of frustrated phagocytosis is suggested by the Stanton hypothesis of an overriding significance of particle dimension for disease induction by durable EPs. A general concept is that EMPs longer than a phagocytic cell's linear dimensions cannot be completely incorporated in a phagosome. Recruitment of membrane from the Golgi apparatus or endoplasmic reticulum may provide extensive addition to the plasma membrane for a cell's attempted invagination to accommodate a long EMP in a phagosomal membrane [Aderem 2002]. However, because of the length of the EMP relative to the dimensions of the cell, the final phagosomal structure is topologically an annulus extending fully through the cell, rather than an enclosed vacuole fully within the cell. Following uptake of nonelongate particles, there is a maturation of the phagosomal membrane; the initial phagosomal membrane is that of the cell's external plasmalemma, which cannot kill or digest phagocytosed material. After sealing of the fully invaginated phagosomal vesicle in the interior of the cell, there is a rapid and extensive change in the membrane composition [Scott et al. 2003]. This involves, in part, an association with lysosomal vesicles and exposure of particles within the secondary phagosome or phagolysosome to lytic enzymes and adjusted pH conditions. It is speculated that failure of the phagosome to close, as occurs in frustrated phagocytosis, induces dysfunction of the system. Conventional phagocytosis of nonelongate particles can lead to a respiratory or oxidative burst of membrane-localized NADPH oxidase of SO radicals, which may be converted to H_2O_2, hydroxyl radicals, and other toxic reactive products of oxygen. If these are released extracellularly in connection with frustrated phagocytosis, they are potentially harmful to the tissue [Bergstrand 1990].

Failure of normal phagocytosis to be completed may affect the duration or intensity of the phagocytic response. It may also affect the generation or release of reactive species or membranolytic digestive enzymes into the still-exterior annulus. Another possible effect is to alter the maturation of the annular frustrated phagocytic membrane from the normal structural and functional evolution of a closed phagolysosomal vesicle fully interior to the cell. Even in the response to such a frustrated phagocytosis, there might be some mineral specificity beyond morphology alone for EMP-induced release of reactive species. Amosite fibers, MMVF, silicon carbide fibers, and RCF-1 fibers all stimulated modest release of SO that was not dose-dependent in isolated rat alveolar macrophages. However, when IgG, a normal component of lung lining fluid, was adsorbed onto the fiber surfaces, SO release was strongly enhanced for all but the silicon carbide fibers. SO release correlated with fibers' IgG-adsorptive capacity, except for the amosite, which required only poorly adsorbed IgG for strong activity, suggesting some mineral specificity beyond morphology alone for the EMP-induced cellular respiratory burst [Hill et al. 1996].

2.9.4.3.4 Cellular Responses to Initiation of Toxicity

Subsequent to initiating damage by direct or induced ROS generation—by direct membranolysis generated by interactions of mineral surface sites with membrane lipids or glycoproteins, or by not-fully-defined toxic response to morphology-based frustrated phagocytosis—a standard model for consequent complex cellular response has evolved and has been the subject of extensive and detailed analyses [Mossman et al. 1997]. EMP-generated primary toxic stimuli to the cell are subject to signal transduction by mitogen-activated protein kinase (MAPK), beginning an intracellular multiple kinase signal cascade, which then induces transcription factors in the nucleus such as activator protein (AP)-1 or nuclear factor kappa beta (NF-κB), which in turn regulate the transcription of mRNA from genes for TNF-α or other cytokines involved in cell proliferation or inflammation.

Fibers of the six asbestos minerals generate MAPK in lung epithelium *in vitro* and *in vivo*, increasing AP-1 transcription activation, cell proliferation, death, differentiation, or inflammation. This is synergistic with cigarette smoke [Mossman et al. 2006]. Macrophage release of oxidants or mitogenic factors through such a pathway could then cause cell proliferation or DNA damage [Driscoll et al. 1998]. In contrast to MMVF-10 and RCF-4, amosite and two other carcinogenic fibers (silicon carbide and RCF-1) produced significant dose-dependent translocation of NF-κB to the nucleus in A549 lung epithelial cells. It was hypothesized that carcinogenic fibers have greater free radical activity, which produces greater oxidative stress and results in greater translocation of NF-κB to the nucleus for the transcription of pro-inflammatory genes (e.g., cytokines) [Brown et al. 1999]. Crocidolite induced AP-1 *in vitro* in JB6 cells and induced AP-1 transactivation in pulmonary and bronchial tissue after intratracheal instillation in transgenic mice, apparently mediated by activation of MAPK [Ding et al. 1999]. Chrysotile challenge to blood monocytes co-cultured with bronchial epithelial cells resulted in elevated levels in epithelial cells of protein-tyrosine kinase activity, NF-κB activity, and mRNA levels for interleukin (IL)-1β, IL-6, and TNF-α. Protein-tyrosine kinase activity, NF-κB activity, and mRNA synthesis were inhibited by antioxidants, suggesting ROS-dependent NF-κB-mediated transcription of inflammatory cytokines in bronchial epithelial cells [Drumm et al. 1999].

Chemokines known to be associated with particle-induced inflammation were found to be secreted by mesothelial cells after amosite challenge to cultured rat pleural mesothelial cells, and they were found in pleural lavage of rats challenged *in vivo* [Hill et al. 2003].

Fibers from crocidolite (asbestiform riebeckite) and EMPs from nonfibrous milled riebeckite increased phosphorylation and activity of a MAPK cascade in association with induction of an inflammatory state of rat pleural mesothelial cells and progenitor cells of malignant mesothelioma. Amelioration by preincubation with vitamin E indicated this to be an oxidative stress effect [Swain et al. 2004]. Lung lysate, cells from bronchoalveolar lavage, and alveolar macrophages and bronchiolar epithelial cells from lung sections from rats exposed to crocidolite or chrysotile fibers contained nitrotyrosine and phosphorylated extracellular signal-regulated kinases (ERKs); nitrotyrosine is a marker for peroxynitrile that activates ERK signaling pathways, altering protein function [Iwagaki 2003]. In vitro challenge of human bronchiolar epithelial cells with crocidolite or chrysotile fibers induced tissue factor (TF) mRNA expression and induced NF-κB and other transcription factors that bind the TF gene promoter. TF *in vivo* is involved in blood coagulation with inflammation and tissue remodeling [Iakhiaev et al. 2004]. Asbestos fibers activate an ERK pathway *in vitro* in mesothelial and epithelial cells. Crocidolite challenge to mice results in phosphorylation of ERK in bronchiolar and alveolar type II epithelial cells, epithelial cell hyperplasia, and fibrotic lesions. Epithelial cell signals through the ERK pathway lead to tissue remodeling and fibrosis [Cummins et al. 2003].

Crocidolite and erionite fibers, but not nonfibrous milled riebeckite, upregulated the expression of epidermal growth factor receptor (EGFR) in rat pleural mesothelial cells *in vitro*. Cell proliferation was co-localized subsequent to EGFR, suggesting initiation of a cell-signaling cascade to cell proliferation and cancer [Faux et al. 2000]. "Long" amosite fibers were more active than "short" amosite fibers in causing (1) damage to nude DNA; (2) *in vitro* cytotoxicity in a human lung epithelial cell line; (3) free radical reactions; (4) inhibition of glycerol-6-phosphate dehydrogenase and pentose phosphate pathways; (5) decrease in intracellular reduced glutathione; (6) increase in thiobarbituric acid reaction substances; and (7) leaking of LDH [Riganti et al. 2003].

An important paradox or seeming failure of *in vitro* studies concerns mesothelioma. Although chrysotile and amphibole asbestos fibers each clearly induce malignant mesothelioma *in vivo*, they do not transform primary human mesothelial cells *in vitro*, whereas erionite fibers do. Asbestos fibers can induce some genotoxic changes; crocidolite fibers induced cytogenotoxic effects, including increased polynucleated cells and formation of 8-OHdG in a phagocytic human mesothelial cell line, but did not induce cytogenotoxic effects in a nonphagocytic human promyelocytic leukemia cell line [Takeuchi et al. 1999]. Tremolite, erionite, RCF-1, and chrysotile fiber challenges of human-hamster hybrid A(L) cells showed chrysotile fibers to be significantly more cytotoxic. Mutagenicity was not seen at the hypoxanthine-guanine phosphoribosyltransferase (HPRT) locus for any of the fibers. Erionite and tremolite fibers induced dose-dependent mutations at the gene marker on the only human chromosome in the hybrid cell. Erionite was the most mutagenic type of fiber. RFC-1 fibers were not mutagenic, in seeming contrast to their known induction of mesothelioma in hamsters [Okayasu et al. 1999]. Crocidolite fibers induced significant but reversible DNA single-strand breaks in transformed human pleural mesothelial cells, and TNF-α induced marginal increases; co-exposure to crocidolite fibers and TNF-α caused greater damage than fibers alone. Antioxidant enzymes did not reduce the DNA damage, suggesting a mechanism of damage other than by free radicals [Ollikainen et al. 1999]. Crocidolite fibers were also very cytotoxic to the cells; presumably, cell death may prevent the observation of cell transformation. In vitro challenge to mesothelial cells and to fibroblast cells by crocidolite fibers, but not by glass wool, induced dose-dependent cytotoxicity and increased DNA synthesis activity [Cardinali et al. 2006]. Crocidolite fibers were found to induce TNF-α secretion and receptors in human mesothelial cells, and TNF-α reduced cytotoxicity of crocidolite fibers by activating NF-κB and thus improving cell survival and permitting expression of cytogenetic activity [Yang et al. 2006]. Erionite fibers transformed immortalized nontumorigenic human mesothelial cells *in vitro* only when exposed in combination with IL-1β or TNF-α [Wang et al. 2004]. Erionite fibers were poorly cytotoxic but induced proliferation signals and a high growth rate in hamster mesothelial cells. Long-term exposure to erionite fibers resulted in transformation of human mesothelial cells *in vitro*, but exposure to asbestos fibers did not transform those cells [Bertino et al. 2007]. In vitro challenge of mesothelial cells to asbestos fibers induced cytotoxicity and apoptosis but not transformation. In vitro challenge of human mesothelial cells to asbestos fibers induced the ferritin heavy chain of iron-binding protein, an anti-apoptotic protein, with decrease in H_2O_2 and other ROS and resistance to apoptosis [Aung et al. 2007]. This was seen also in a human malignant mesothelial cell line.

The question of a co-carcinogenic effect of asbestos fibers with a virus has been raised. Most malignant mesotheliomas are associated with asbestos exposures, but only a fraction of exposed individuals develop mesothelioma, indicating that other factors may play a role. It has been suggested that simian virus 40 (SV40) and asbestos fibers may be co-carcinogens. SV40 is a DNA tumor virus that causes mesothelioma in hamsters and has been detected in several human mesotheliomas. Asbestos fibers appear to increase SV40-mediated transformation of human mesothelial cells *in vitro* [Carbone et al. 2002]. In an *in vivo* demonstration of co-carcinogenicity of SV40 and asbestos fibers, mice containing high copy numbers of SV40 viral oncogene rapidly developed fast-growing mesothelioma following asbestos challenge. Transgenic copy number was proportional to cell survival and *in vitro* proliferation [Robinson et al. 2006].

Various mechanisms protect cells and tissues against oxidants, and it is conceivable that genetic and acquired variations in these systems may account for individual variation in the response to oxidative stress [Driscoll et al. 2002]. Similarly, species differences in antioxidant defenses or the capacity of various defenses may underlie differences in response to xenobiotics that act, in whole or part, through oxidative mechanisms. Oxidative mechanisms of response to xenobiotics are especially relevant to the respiratory tract, which is directly and continually exposed to an external environment containing oxidant pollutants (e.g., ozone, oxides of nitrogen) and particles that may generate oxidants as a result of chemical properties or by stimulating production of cell-derived oxidants. In addition, exposure to particles or other pollutants may produce oxidative stress in the lung by stimulating the recruitment of inflammatory cells. For example, the toxicity of asbestos fibers likely involves the production of oxidants, such as hydroxyl radical, SO, and H_2O_2. Studies have also shown that asbestos fibers and other mineral particles may act by stimulating cellular production of ROS and reactive nitrogen species. In addition to direct oxidant production, exposure to asbestos and SVFs used in high-dose animal studies stimulates the recruitment and activation of macrophages and polymorphonuclear leukocytes that can produce ROS through the activity of NADPH oxidase on their cell membranes. Developing an understanding of the oxidative stress/NF-κB pathway for EMP-mediated inflammation and the interplay among exposure-induced oxidant production, host antioxidant defenses, and interindividual or species variability in defenses may be very important for appropriate risk assessments of inhaled EMPs [Donaldson and Tran 2002].

2.9.4.4 Studies Comparing EMPs from Amphiboles with Asbestiform versus Nonasbestiform Habits

Smith et al. [1979] compared tumor induction after IP injection in hamsters of two asbestiform tremolites, two nonasbestiform prismatic tremolitic talcs, and one tremolitic talc of uncertain asbestiform status. No tumors were observed following the nonasbestiform tremolite challenge, in contrast to the asbestiform tremolites. However, tumors were observed from the tremolitic talc of uncertain amphibole status. In rule-making, OSHA [1992] noted the small number of animals in the study, the early death of many animals, and the lack of systematic characterization of particle size and aspect ratio. Subsequent analyses (by chemical composition) performed on the nonasbestiform tremolitic talc from the study, which was not associated with mesothelioma, found 13% of particles had at least a 3:1 aspect ratio [Wylie et al. 1993]. A prismatic, nonasbestiform tremolitic talc and an asbestiform tremolite from the study were analyzed for aspect ratio [Campbell et al. 1979]. The researchers analyzed 200 particles of the asbestiform tremolite sample and found 17% had an aspect ratio of 3:1 or greater and 9.5% had an aspect ratio greater than 10:1. Analysis of 200 particles of the prismatic tremolite showed that 2.5% (five EMPs) had an aspect ratio of 3:1 or greater and 0.5% (one EMP) had an aspect ratio greater than 10:1.

Wagner et al. [1982] challenged rats by IP injection of tremolite asbestos, a prismatic nonasbestiform tremolite, or a tremolitic talc considered nonasbestiform containing a limited number of long fibers. Only the tremolite asbestos produced tumors; mesothelioma was found in 14 of 37 animals. The authors speculated that tumor rate may have risen further if the testing period had not been shortened due to infection-induced mortality. Per microgram of injected dose, the asbestiform sample contained 3.3×10^4 nonfibrous particles, 15.5×10^4 fibers, and 56.1×10^3 fibers >8 μm long and <1.5 μm wide. Corresponding values for the prismatic amphibole were 20.7×10^4, 4.8×10^4, and 0. Tremolitic talc values were 6.9×10^4, 5.1×10^4, and 1.7×10^3. Infection-reduced survival prevented evaluation of a crocidolite-exposed positive control.

Another IP injection study with the rat used six samples of tremolite of different morphological types [Davis et al. 1991]. For three asbestiform samples, mesothelioma occurred in 100%, 97%, and 97% of exposed animals, at corresponding doses of 13.4×10^9 fibers with aspect ratio >3:1 per 121×10^6 fibers with length >8 μm and diameter <0.25 μm (100% of exposed animals); 2.1×10^9 per 8×10^6 (97% of exposed animals); and 7.8×10^9 per 48×10^6 (97% of exposed animals). For an Italian tremolite from a nonasbestos source and containing relatively few asbestiform fibers (1.0×10^9 per 1×10^6), mesothelioma was found in two-thirds of the animals, with delayed expression. For two nonasbestiform tremolites (0.9×10^9 per 0; 0.4×10^9 per 0), tumors were found in 12% and 5% of the animals, respectively; of these, and the response rate of 12% was above the expected background rate. The Italian sample resulting in 67% mesothelioma incidence contained only one-third the number of EMPs >8 µm long, in comparison with the nonasbestiform sample associated with 12% mesothelioma, and those two samples contained an approximately equal number of fibers with length >8 µm and width <0.5 µm. The preparations of the three asbestiform samples and the Italian sample were essentially identical. However, the two nonasbestiform samples associated with low mesothelioma incidence required significantly different pretreatment: the first required multiple washings and sedimentation, and the second, grinding under water in a micronizing mill. It was noted that those two nonasbestiform samples and the Italian sample contained minor components of long, thin asbestiform tremolite fibers. This study suggested that carcinogenicity may not depend simply on the number of EMPs, and the researchers called for methods of distinguishing "carcinogenic tremolite fibers" from nonfibrous tremolite dusts that contain similar numbers of EMPs of similar aspect ratios [Davis et al. 1991]. It has been suggested that the response observed for the Italian tremolite is of a pattern expected for a low dose of highly carcinogenic asbestos tremolite [Addison 2007].

A recent review of past studies of varieties of tremolite and the limitations of earlier studies (e.g., their use of injection or implantation versus inhalation) suggested that, on the basis of observed differences in the carcinogenicity of tremolite asbestos and nonasbestiform prismatic tremolite, differences in carcinogenicity of amphibole asbestos fibers and nonasbestiform amphibole cleavage fragments are sufficiently large to be discernable, even with the study limitations [Addison and McConnell 2008]. The authors also concluded the evidence supports a view that shorter, thicker cleavage fragments of the nonasbestiform amphiboles are less hazardous than the thinner asbestos fibers [Addison and McConnell 2008].

In summary, several types of animal studies have been conducted to assess the carcinogenicity and fibrogenicity of asbestiform and nonasbestiform tremolite fibers and other EMPs. Tremolite asbestos was found to be both fibrogenic and carcinogenic in rats by inhalation. However, the data for other particle forms of tremolite and for other amphiboles in general are much more limited and are based primarily on mesotheliomas produced by intrapleural administration studies in rats. Although these tests can be highly sensitive for specific material-tissue interactions, these studies bypass the lung entirely and thus provide no information on the test material's potential for causing lung tumors. In addition, they have often been criticized for employing a nonphysiological route of administration. Some of the older studies [Smith et al. 1979; Wagner et al. 1982] are difficult to interpret because of inadequate characterization of the tremolite preparation that was used, although the studies do show fewer or no tumors from prismatic tremolite than from asbestiform tremolite when tested similarly. Unfortunately, doses used in most animal studies are generally reported in terms of mass (e.g., 10, 25, or 40 mg/rat). Unless the test preparations are well characterized in terms of fiber counts and fiber size distributions, it is difficult to relate the mass-based dose in the animals to fiber count concentrations used to assess human occupational exposures. Where semiquantitative fiber count and size distribution data are provided, as in the study report by Davis et al. [1991], it is evident that the prismatic tremolite samples contain fewer particles >5 µm in length with an aspect ratio of >3:1 per 10-mg dose than the asbestiform tremolite samples, and the median particle widths are greater for the prismatic tremolite samples. Although the prismatic tremolite samples clearly generated fewer mesotheliomas than the asbestiform tremolite samples, it is not apparent whether the tumorigenic potency per EMP is lower for the nonasbestiform tremolites.

In a comparison of one asbestiform and two nonasbestiform tremolites, cellular *in vitro* assays for LDH release, beta-glucuronidase release, cytotoxicity, and giant cell formation showed relative toxicities that parallel the differences seen in an *in vivo* rat IP injection study of tumorigenicity using the same samples [Wagner et al. 1982]. In vitro cellular or organ tissue culture studies showed squamous metaplasia and increased DNA synthesis in tracheal explant cultures treated with long glass fibers or with crocidolite or chrysotile fibers, whereas cleavage fragments from their nonasbestiform analogues, riebeckite and antigorite, were not active [Woodworth et al. 1983]. For alveolar macrophages *in vitro*, crocidolite fibers induced the release of ROS at an order of magnitude greater than cleavage fragments from nonasbestiform riebeckite [Hansen and Mossman 1987]. Similar differences were observed in hamster tracheal cells for

- induction of ornithine decarboxylase, an enzyme associated with mouse skin cell proliferation and tumor promotion [Marsh and Mossman 1988];
- stimulating survival or proliferation in a colony-forming assay using those hamster tracheal epithelial cells [Sesko and Mossman 1989];
- activation of proto-oncogenes in tracheal epithelial and pleural mesothelial cells *in vitro* [Janssen et al. 1994]; and
- cytotoxicity [Mossman and Sesko 1990].

A recent review concludes that a large body of work shows that cleavage fragments of amphiboles are less potent than asbestos fibers in a number of *in vitro* bioassays [Mossman 2008].

Several studies have investigated induction of molecular pathways in cells exposed to EMPs. The goals of such studies were to elucidate mechanisms involved in fiber-induced pathogenicity and to compare the relative potency of asbestiform and nonasbestiform EMPs. However, seemingly contradictory findings between some experiments suggest that improved methods for preparation of size-selected test particles and more extensive characterization of EMPs are needed.

Although few animal studies of nonasbestiform amphibole dusts have been conducted, this research has revealed generally significant differences in pathogenicity between nonasbestiform and asbestiform amphiboles. There are few findings of biological effects or tumorigenicity induced by samples classified as nonasbestiform, and there are rational hypotheses as to the cause of those effects. There are general fundamental uncertainties concerning EMP properties and biological mechanisms that determine mineral particle toxicities and pathogenicities, specifically concerning the similarities or differences in disease mechanisms between EMPs from asbestiform versus nonasbestiform amphiboles. In vitro studies have generally shown differences in specific toxic activities between some asbestiform and nonasbestiform amphibole EMPs, although *in vitro* systems are not yet able to predict relative pathogenic risk for mineral fibers and other EMPs. This suggests a focus of research to determine if and when nonasbestiform amphibole EMPs are active for tumorigenicity or other pathology, if there is a threshold for those activities, and if distinguishing conditions or properties that determine such pathogenicity can be found.

2.9.5 Thresholds

Discussions of thresholds for adverse health effects associated with exposure to asbestos fibers and related EMPs have focused on the characteristics of dimension, including length, width, and the derived aspect ratio, as well as concentration. Although other particle characteristics discussed above may impact these thresholds or may have thresholds of their own that impact the toxicity of EMPs, they are not well discussed in the literature. The following discussion is focused on thresholds for dimension and concentration.

The seminal work of Stanton et al. [1981] laid the foundation for much of the information on dimensional thresholds. Their analyses showed that malignant neoplasms in exposed rats were best predicted by the number of EMPs longer than 8 µm and thinner than 0.25 µm. However, the number of EMPs in other size categories, having lengths greater than 4 µm and widths up to 1.5 µm, were also highly correlated with malignant neoplasms. Also, some samples with relatively larger proportions of shorter particles, such as the tremolites, produced high rates of tumors. Lippmann [1988, 1990] reviewed the literature and suggested that lung cancer is most closely associated with asbestos fibers longer than 10 µm and thicker than 0.15 µm, whereas mesothelioma is most closely associated with asbestos fibers longer than 5 µm and thinner than 0.1 µm. Evidence from animal studies and some *in vitro* studies suggests that short asbestos fibers (e.g., <5 µm long) may play a role in fibrosis but are of lesser concern than longer asbestos fibers for cancer development.

Berman et al. [1995] statistically analyzed aggregate data from 13 inhalation studies in which rats were exposed to 9 types of asbestos (4 chrysotiles, 3 amosites, 1 crocidolite, and 1 tremolite asbestos) to assess fiber dimension and mineralogy as predictors of lung tumor and mesothelioma risks. Archived samples from the studies were reanalyzed to provide detailed information on each asbestos structure, including mineralogy (chrysotile, amosite, crocidolite, or tremolite), size (length and width, each in 5 categories), type (fiber, bundle, cluster, or matrix), and complexity (number of identifiable components of a cluster or matrix). Multiple concentrations (each for asbestos structures with different specified characteristics) were calculated for the experimental exposures. Although no univariate index of exposure adequately described lung tumor incidence observed across all inhalation studies, certain multivariate indices of exposure did adequately describe outcomes. Fibers and bundles longer than 5 µm and thinner than 0.4 µm contributed to lung tumor risk; very long (>40 µm) and very thick (>5 µm) complex clusters and matrices possibly contributed. Although structures <5 µm long did not contribute to lung tumor risk, potency of thin (<0.4 µm) structures increased with increasing length above 5 µm, and structures >40 µm long were estimated to be about 500 times more potent than structures between 5 and 40 µm long. With respect to lung tumor risk, no difference was observed between chrysotile and amphibole asbestos. With respect to mesothelioma risk, chrysotile was found to be less potent than amphibole asbestos. Although the analysis of Berman et al. [1995] was limited to studies of asbestos exposure, similar statistical approaches may be adaptable to assess study outcomes from exposures to a broader range of EMPs, beyond asbestos.

In addressing the issue of a length threshold, the Health Effects Institute [HEI 1991] concluded that asbestos fibers <5 μm long appear to have much less carcinogenic activity than longer fibers and may be relatively inactive. A panel convened by ATSDR [2003] concluded that "given findings from epidemiological studies, laboratory animal studies, and *in vitro* genotoxicity studies, combined with the lung's ability to clear short fibers, the panelists agreed that there is a strong weight of evidence that asbestos and SVFs shorter than 5 μm are unlikely to cause cancer in humans." Also, an EPA [2003] peer consultant panel "agreed that the available data suggest that the risk for fibers <5 μm long is very low and could be zero." They generally agreed that the width cutoff should be between 0.5 and 1.5 μm but deserved further analysis.

However, Dodson et al. [2003] have argued that it is difficult to rule out the involvement of short (<5 μm) asbestos fibers in causing disease, because exposures to asbestos fibers overwhelmingly involve fibers shorter than 5 μm, and fibers observed in the lung and in extrapulmonary locations are also overwhelmingly shorter than 5 μm. For example, in a study of malignant mesothelioma cases, Suzuki and Yuen [2002] and Suzuki et al. [2005] found that the majority of asbestos fibers in lung and mesothelial tissues were shorter than 5 μm. Another study [Berman and Crump 2008a] showed that the best estimate for the potency of asbestos fibers between 5 and 10 μm is zero, but the hypothesis that these shorter fibers are not potent could not be rejected. Likewise, in no case could the hypothesis be rejected that fibers shorter than 5 μm are not potent.

NIOSH investigators have recently evaluated the relationship between the dimensions (i.e., length and width) of airborne chrysotile fibers and risks for developing lung cancer or asbestosis, by updating the cohort of chrysotile-exposed textile workers previously studied by Dement et al. [1994], Stayner et al. [1997], and Hein et al. [2007]. Archived air samples collected at this chrysotile textile plant were reanalyzed by TEM to generate exposure estimates based on bivariate fiber-size distribution [Dement et al. 2008]. TEM analysis of sampled fibers found all size-specific categories (35 categories were assigned, based on combinations of fiber width and length) to be highly statistically significant predictors of lung cancer and asbestosis [Stayner et al. 2007]. The smallest fiber size–specific category was thinner than 0.25 μm and <1.5 μm long. The largest size-specific category was thicker than 3.0 μm and >40 μm long. Both lung cancer and asbestosis were most strongly associated with exposures to thin fibers (<0.25 μm), and longer fibers (>10 μm) were found to be the strongest predictors of lung cancer. A limitation of the study is that cumulative exposures for the cohort were highly correlated across all fiber-size categories, which complicates the interpretation of the study results.

In addition to length and width, an important parameter used to define EMPs is the aspect ratio. The use of the 3:1 length:width aspect ratio as the minimum to define an EMP was not established on scientific bases such as toxicity or exposure potential. Rather, the decision was based on the ability of the microscopist to determine the elongate nature of a particle [Holmes 1965], and the practice has been carried through to this day. As bivariate analyses are conducted, the impact of aspect ratio, in addition to length and width, on toxicity and health outcomes needs to be addressed.

As discussed in Section 2.4.2, the nature of occupational exposures to asbestos has changed over the last several decades. Once dominated by chronic exposures in textile mills, friction product manufacturing, and cement pipe fabrication, current occupational exposures to asbestos in the United States are primarily occurring during maintenance activities or remediation of buildings containing asbestos. These current occupational exposure scenarios frequently involve short-term, intermittent exposures. The generally lower current exposures give added significance to the question of whether or not there is an asbestos exposure threshold below which workers would incur no risk of adverse health outcomes.

Risk assessments of workers occupationally exposed to asbestos were reviewed by investigators sponsored by the Health Effects Institute [1991]. They found that dose-specific risk is highly dependent on how the measurement of dose (exposure) was determined. A common problem with many of the epidemiological studies of workers exposed to asbestos is the quality of the exposure data. Few studies have good historical exposure data, and the available data are mostly area samples with concentrations reported as millions of particles per cubic foot of air (mppcf). Although correction factors have been used to convert exposures measured in mppcf to f/cm^3, the conversions were often based on more recent exposure measurements collected at concentrations lower than those prevalent in earlier years. In addition, a single conversion factor was typically used to estimate exposures throughout a facility, which may not accurately represent differences in particle sizes and counts at different processes in the facility.

More recently, the concept of a concentration threshold has been reviewed by Hodgson and Darnton [2000]. It is generally accepted that lung fibrosis requires relatively heavy exposure to asbestos and that the carcinogenic response of the lung may be an extension of the same inflammatory processes that produce lung fibrosis. Some evidence for a threshold is provided by an analysis of a chrysotile-exposed cohort, which suggests a potential threshold dose of about 30 f/mL-yr to produce radiologically evident fibrosis [Weill 1994]. Another study of necropsy material from textile workers exposed to chrysotile shows a distinct step increase in fibrosis for exposures in the 20–30 f/mL-yr range [Green et al. 1997]. However, a study of textile mill workers exposed to chrysotile did not reveal evidence of significant concentration thresholds for either asbestosis or lung cancer [Stayner et al. 1997]. Hodgson and Darnton [2000] pointed out that any evidence suggesting a threshold for chrysotile would likely not apply to amphibole asbestos because radiologically evident fibrosis has been documented among workers exposed to amphibole asbestos at low levels (<5 f/mL-yr). They concluded that if a concentration threshold exists for amphiboles, it is very low.

For mesothelioma, Hodgson and Darnton [2000] identified cohorts with high rates of mesothelioma at levels of exposure below those at which increased lung cancer has been identified; in some studies, the proportion of mesothelioma cases with no likely asbestos exposure is much higher than expected. Hodgson and Darnton [2000] concluded that these studies support a nonzero risk, even from brief, low-level exposures.

Animal studies using intraperitoneal and intrapleural injection of asbestos fibers cited by Ilgren and Browne [1991] suggest a possible threshold concentration for mesothelioma. However, it is not clear how this would be useful to determine a threshold for inhalation exposure in humans.

2.10 Analytical Methods

Available analytical methods can characterize the size, morphology, elemental composition, crystal structure, and surface composition of bulk materials and individual airborne particles. There are two separate paradigms for selecting among these methods for their use or further development for application to EMPs: one is for their support of standardized surveys or compliance assessments of workplace exposures to EMPs; another is for their support of research to identify physicochemical properties of EMPs that are critical to predicting toxicity or pathogenic potential for lung fibrosis, lung cancer, or mesothelioma. The former refers to analytical methods that can be applied to samples of airborne particles, whereas the latter can be used to characterize airborne particles and bulk materials.

Cost, time, availability, standardization requirements, and other pragmatic factors limit the selection of analytical methods for standardized analysis of field samples for the first set of uses. Additionally, those uses require methods employed historically to establish exposure-response relationships. Principal among these analyses for standardized industrial hygiene use is an optical microscopy method, PCM (e.g., NIOSH Method 7400 or equivalent) [NIOSH 1994a]. Under the current NIOSH REL for airborne asbestos fibers, particles are counted if they are EMPs (i.e., mineral particles with an aspect ratio [length:width] of 3:1 or greater) of the covered minerals and they are longer than 5 µm when viewed microscopically according to NIOSH Method 7400 or its equivalent. The assumption when using this method is that all particles meeting the dimensional criteria are airborne asbestos fibers, because PCM cannot identify the chemistry or crystalline structure of a particle. This assumption may be appropriate in situations where the majority of particles are reasonably assumed to be one of the minerals covered by the airborne asbestos fiber REL. Electron microscopy can be used to determine the actual proportion of the total particles that are covered by the airborne asbestos fiber REL, and this proportion can be used to adjust the count from PCM (NIOSH Method 7402) [NIOSH 1994b]. Such counts are known as PCM-equivalents, or PCMe. Note that this is not the same procedure as counting particles that would meet the PCM criteria under the electron microscope. Methods have been developed for performing counts under both scanning electron microscopy (SEM) [ISO 2002] and TEM [40 CFR[§§] 763.A.E (2001)]. However, only a few countries (e.g., Germany, Austria, Netherlands, and Switzerland) use SEM routinely for counting.

Characterization of bulk minerals is a process known as petrographic analysis. Petrographic analysis includes a number of techniques, including polarized light microscopy (PLM), electron microscopy (SEM or TEM), X-ray diffraction (XRD), X-ray fluorescence (XRF), and electron microprobe analysis. Other techniques, such as infrared and Raman spectroscopy and surface area measurements, can also be used. Some of these techniques can also be applied to individual airborne particles. If it is determined that the toxicity of EMPs has a basis in properties that can be measured by one or more of these techniques, then it may be possible to tailor analytical procedures in the future to more precisely estimate risk.

Care should be taken in developing or applying new analytical methods to the analysis of asbestos for standardized and compliance assessments. The use of new or different analytical methods to assess exposures must be carefully evaluated and validated to ensure that they measure exposures covered by the health protection standard.

[§§]*Code of Federal Regulations.* See CFR in references.

One consideration in developing new methods for assessing workplace exposures is that silicate mineral particles thinner than the resolution of PCM in NIOSH Method 7400 are in the same size range as the deposition minimum observed for small particles in human respirable particle studies. Current standards for assessing particle dose are based on particle penetration into the human respiratory system, which may overestimate deposition [ISO 1995a]. More recently, proposals have been developed to account for deposition [Vincent 2005]. In addition, after depositing in the lung, a single asbestos fiber bundle can split into a great many individual fibrils. Accounting for this behavior may be important in developing methods that accurately represent risk from exposure.

It is also important to recognize that the sampling and analytical methods for assessing workplace exposures to EMPs have different constraints from methods used to assess environmental exposures. NIOSH is focused on developing and validating methods for assessing workplace exposures to EMPs and provides assistance in developing environmental exposure methods, where possible and appropriate, through its relationships with other federal agencies.

2.10.1 NIOSH Sampling and Analytical Methods for Standardized Industrial Hygiene Surveys

The analytical components of NIOSH's REL for asbestos exposure take on substantial significance because the current REL was set on the basis of the limit of quantification (LOQ) of the PCM method with use of a 400-L sample, rather than solely on estimates of the health risk. Had a lower LOQ been possible, a lower REL may have been proposed to further reduce the risk of occupational cancer among asbestos-exposed workers. With advances in sampling pump capabilities—and if interference from other dust is not excessive—a lower limit of quantitation and thus a shorter averaging time for the REL may be possible.

PCM was designated as the principal analytical method for applying the REL because it was thought to be the most practical and reliable available method, particularly for field assessments. The particle counting rules specified for PCM analysis of air samples result in an index of exposure that has been used with human health data for risk assessment. PCM-based counts do not enumerate all EMPs because very thin particles, such as asbestos fibrils, are typically not visible by PCM when using NIOSH Method 7400. The ratio of countable EPs to the total number of EPs collected on air samples can therefore vary for samples collected within the same workplace, as well as between different workplaces where the same or different asbestos materials are handled [Dement and Wallingford 1990]. The result of this is that equivalent PCM asbestos exposure concentrations determined at different workplaces would be considered to pose the same health risk, when, in fact, those risks may be different because of unknown amounts of unobserved fibers on the samples. It is commonly stated that particles thinner than about 0.2–0.25 µm typically cannot be observed with PCM because they are below the resolution limits of the microscope. However, the results for PCM counts may also vary according to the index of refraction of the material being examined. When the index of refraction of the particle is similar to that of the filter substrate or mounting medium, the ability to resolve particles is less than when the refractive index of the particle differs from that of the substrate [Kenny and Rood 1987]. When a microscope is calibrated appropriately for NIOSH Method 7400 and triacetin is used as the mounting medium, calculation and experiment have indicated that chrysotile fibers as thin as 0.15 μm can be resolved [Rooker et al. 1982], which implies that amphibole fibers thinner than 0.2 μm and with higher refractive index may actually be visible and potentially counted.

Individual asbestos fibrils range in width from <10 nm (0.01 μm) for chrysotile up to 40 nm (0.04 μm) or more for amosite. Thus, individual asbestos fibrils are not likely to be visible by PCM. Early studies suggested that asbestos particles of 3:1 aspect ratio and longer than 5 μm are not usually individual fibrils but fibrillar bundles that are much wider than fibrils [Hwang and Gibbs 1981; further data cited in Walton 1982], so that the number of particles meeting these criteria counted under PCM were not generally found to differ greatly from the number of particles meeting the same criteria counted under the electron microscope [Lynch et al. 1970; Hwang and Gibbs 1981; Marconi et al. 1984; Dement and Wallingford 1990]. However, more recent studies suggest there is a substantial difference between counts of particles seen by PCM and TEM [Dement et al. 2009], and the importance of these differences in developing new methods to be used routinely for industrial hygiene surveys will be important to reconcile with the particle characteristics important in disease causation.

Another aspect of NIOSH Method 7400 is that two sets of counting rules are specified, depending on the type of fiber analysis. The rules for counting particles for asbestos determination, referred to as the "A" rules, instruct the microscopist to count EPs of any width that are longer than 5 μm and have an aspect ratio of at least 3:1. However, EPs wider than 3 μm are not likely to reach the thoracic region of the lung when inhaled. The "B" counting rules, which are used to evaluate airborne exposure to other EPs, specify that only EPs thinner than 3 μm and longer than 5 μm should be counted [NIOSH 1994a]. The European Union is moving toward a standardized PCM method for evaluating asbestos exposures with counting rules recommended by the World Health Organization (WHO), which specify counting only EPs thinner than 3 μm and with a 3:1 or larger aspect ratio [WHO 1997; European Parliament and Council 2003].

2.10.2 Analytical Methods for Research

For research purposes, it may be important for a more expansive set of analyses to be considered. However, EMPs thinner than the limit of spatial resolution of the optical microscope are thought to be important etiologic agents for disease, so other detection and measurement methods may be needed for improved investigations of the relationship between fiber dimension and disease outcomes.

TEM has much greater resolving power than optical microscopy, on the order of 0.001 μm. Additionally, TEM has the ability to semiquantitatively determine elemental composition by means of EDS. Incident electrons excite electronic states of atoms of the sample, and the atoms decay that excess energy either by emitting an X-ray of frequency specific to the element (X-ray spectroscopy) or by releasing a secondary electron with equivalent kinetic energy (an Auger electron). Furthermore, TEM can provide some level of electron diffraction (ED) analysis of particle mineralogy by producing a mineral-specific diffraction pattern based on the regular arrangement of the particle's crystal structure [Egerton 2005].

The greater spatial resolving power and the crystallographic analysis abilities of TEM and TEM-ED are used in some cases for standardized workplace industrial hygiene characterizations. TEM methods (e.g., NIOSH 7402) are used to complement PCM in cases where there is apparent ambiguity in EMP identification [NIOSH 1994b] and, under the Asbestos Hazardous Emergency Response Act of 1986, the EPA requires that TEM analysis be used to ensure the effective removal of asbestos from schools [EPA 1987]. Each of these methods employs specific criteria for defining and counting visualized fibers and reporting different fiber counts for a given sample. These data subsequently can be independently interpreted according to different definitional criteria, such as those developed by the International Organization for Standardization (ISO), which provides methods ISO 10312 and ISO 13794 [ISO 1995b, 1999].

Improved analytical methods that have become widely available should be reevaluated for complementary research applications or for ease of applicability to field samples. SEM is now generally available in research laboratories and commercial analytical service laboratories. SEM resolution is on the order of ten times that of optical microscopy, and newly commercial field emission SEM (FESEM) can improve this resolution to about 0.01 μm or better, near that of TEM. SEM-EDS and SEM- wavelength dispersive spectrometers (WDS) can identify the elemental composition of particles. It is not clear that SEM-backscatter electron diffraction analysis can be adapted to crystallographic analyses equivalent to TEM-ED capability. The ease of sample collection and preparation for SEM analysis in comparison with TEM and some of the advantages of SEM in visualizing fields of EMPs and EMP morphology suggest that SEM methods should be reevaluated for EMP analyses, both of field samples and for research [Goldstein 2003].

Research on mechanisms of EMP toxicity includes concerns for surface-associated factors. To support this research, elemental surface analyses can be performed by scanning Auger spectroscopy on individual particles with widths near the upper end of SEM resolution. In scanning Auger spectroscopy, the Auger electrons stimulated by an incident electron beam are detected; the energy of these secondary electrons is low, which permits only secondary electrons from near-surface atoms to escape and be analyzed, thus analyzing the particle elemental composition to a depth of only one or a few atomic layers [Egerton 2005]. This method has been used in some pertinent research studies (e.g., assessing effects on toxicity of leaching Mg from chrysotile fiber surfaces) [Keane et al. 1999]. Currently, this form of analysis is time-consuming and not ideal for the routine analysis of samples collected in field studies.

Surface elemental composition and limited valence state information on surface-borne elements can be obtained by X-ray photoelectron spectroscopy (XPS or ESCA), albeit not for individual particles. XPS uses X-ray excitation of the sample, rather than electron excitation as used in SEM-EDS or TEM-EDS. The X-rays excite sample atom electrons to higher energy states, which then can decay by emission of photoelectrons. XPS detects these element-specific photoelectron energies, which are weak and therefore emitted only near the sample surface, similar to the case of Auger electron surface spectroscopy. In contrast to scanning Auger spectroscopy, XPS can in some cases provide not only elemental but also valence state information on atoms near the sample surface. However, in XPS the exciting X-rays cannot be finely focused on individual particles, so analysis is made of an area larger than a single particle [Watts and Wolstenholme 2003]. Thus, analysis of a mixed-composition dust sample would be confounded, so XPS is applicable only to some selected or prepared homogeneous materials or to pure field samples.

2.10.3 Differential Counting and Other Proposed Analytical Approaches for Differentiating EMPs

In assessments of asbestos fiber counts in mixed exposures, the use of PCM to determine concentrations of airborne fibers from asbestos minerals cannot ensure that EMPs from nonasbestiform minerals are excluded. No reliable, reproducible methods are available for analysis of air samples that would distinguish between asbestos fibers and EMPs from nonasbestiform analogs of the asbestos minerals. The lack of such validated analytical methods is clearly a major limitation in applying federal agencies' definitions of airborne asbestos fibers.

A technique referred to as differential counting, suggested as an approach to differentiate between asbestiform and nonasbestiform EMPs, is mentioned in a nonmandatory appendix to the OSHA asbestos standard. That appendix points out that the differential counting technique requires "a great deal of experience" and is "discouraged unless legally necessary." It relies heavily on subjective judgment and does not appear to be commonly used except for samples from mines. In this technique, EMPs that the microscopist judges to be nonasbestiform (e.g., having the appearance of cleavage fragments) are not counted; any EMPs not clearly distinguishable as either asbestos or not asbestos in differential counting are to be counted as asbestos fibers. One effect of using differential counting is to introduce an additional source of variability in the particle counts, caused by different "reading" tendencies between microscopists. The technique has not been formally validated and has not been recommended by NIOSH.

For counting airborne asbestos fibers in mines and quarries, ASTM has proposed "discriminatory counting" that incorporates the concepts of differential counting. The proposed method uses PCM and TEM in a tiered scheme. Air samples are first analyzed by PCM. If the initial PCM fiber count exceeds the MSHA permissible exposure limit (PEL), TEM is performed to determine an equivalent PCM count of regulated asbestos fibers only. If the initial PCM count is greater than one-half the PEL but less than the PEL, discriminatory counting is then performed. Discriminatory counts are restricted to fiber bundles, fibers longer than 10 μm, and fibers thinner than 1.0 μm. If the discriminatory count is at least 50% of the initial PCM fiber count, TEM is performed to determine an equivalent PCM count of regulated asbestos fibers only. These results are then compared to regulatory limits [ASTM 2006].

ASTM has begun an interlaboratory study (ILS #282) to determine the interlaboratory precision of "binning" fibers into different classes based on morphology [Harper et al. 2007]. The first part of the validation process was to evaluate samples of ground massive or coarsely crystalline amphiboles and air samples from a taconite mine that have amphibole particulates, of which the majority are characterized as cleavage fragments. Almost none of the observed particles met the Class 1 criteria (i.e., potentially asbestiform on the basis of curved particles and/or fibril bundles). Many particles were classified as Class 2 (i.e., potentially asbestiform on the basis of length >10 μm or width <1 μm), although their morphology suggested they were more likely cleavage fragments. With use of alternate criteria for Class 2 (length >10 μm plus width <1 μm), the number of Class 2 particles was greatly reduced. However, evidence from the literature [Dement et al. 1976; Griffis et al. 1983; Wylie et al. 1985; Siegrist and Wylie 1980; Beckett and Jarvis 1979; Myojo 1999] indicates that as many as 50% of airborne asbestos fibers are <10 μm long. The proportion of asbestos fibers in the length "bin" bracketed by 5 μm and 10 μm was also quite large (about 30%), and the adoption of the alternate Class 2 criteria as length >10 μm and width <1 μm would cause this proportion of asbestos fibers to be classified as nonasbestiform and excluded from counts of asbestos fibers [Harper et al. 2008].

Other procedures have been suggested with the intent of ensuring that the counts on air samples do not include cleavage fragments [IMA-NA 2005; NSSGA 2005]. These procedures include reviewing available geological information and/or results from analysis of bulk materials to establish that asbestos is present in the sampled environment, or specifying dimensional criteria to establish that airborne particulates have population characteristics typical of asbestos fibers (e.g., mean particle aspect ratios exceeding 20:1).

For research purposes, it is critically important that an analytical method that clearly distinguishes between asbestiform and nonasbestiform EMPs be developed, validated, and used. Whether any of these suggested procedures would ensure adequate health protection for exposed workers is unclear, and the practical issues associated with implementing these supplemental procedures are also undetermined.

2.11 Summary of Key Issues

For asbestos fibers and other EMPs, an important question that remains unanswered is "What are the important dimensional and physicochemical determinants of pathogenicity?" Evidence from epidemiological and animal studies indicates that risks for asbestosis and lung cancer decrease with decreasing exposure concentration and that the potency of asbestos is reduced as the fiber length decreases. However, the results from lung burden studies indicate the presence of short asbestos fibers at disease sites, and positive correlations between lung cancer and exposure to short asbestos fibers make it difficult to rule out a role for short asbestos fibers in causing disease.

Other areas of debate remain. For example, there is a continuing debate as to whether or not amphibole asbestos fibers are more potent lung carcinogens than chrysotile fibers. This possible greater potency has been postulated to result from slower dissolution (in lung, interstitial, and phagolysosomal fluids) of amphibole asbestos fibers compared to chrysotile fibers. Thus, amphibole asbestos fibers may tend to persist for longer periods in the lungs and other tissues, thereby imparting a greater potential to trigger carcinogenesis. A related issue that continues to be debated is the magnitude of the difference in potency for mesothelioma associated with types of asbestos fiber.

Understanding the determinants of toxicity of asbestos fibers and other EMPs, as well as of nonelongate mineral particles such as quartz, may help to elucidate some of these issues. The human, animal, and *in vitro* studies performed to date have been on asbestos fibers and a limited number of nonasbestiform EMPs relative to the large and highly variable group of minerals to which there is potential exposure. Additional data are needed to more fully assess the particle characteristics that cause adverse health outcomes and to develop risk assessments.

Information on occupational exposures to nonasbestiform EMPs is limited, as is information on exposure to asbestos fibers and other EMPs in mixed-dust environments, making it difficult to assess the range of particle characteristics, including dimension, in occupational settings with exposures to nonasbestiform EMPs. The few studies that have assessed biopersistence or durability suggest that nonasbestiform EMPs are not as biopersistent as asbestiform fibers of the same length, but more information is needed to systematically assess the ranges and importance of biopersistence in determining toxicity. Any assessment of risk needs to address the influence of EMP dimensions, so studies that systematically compare, to the extent possible, the effects of asbestiform and nonasbestiform particles of similar dimensions and dosimetrics—such as particle count, disaggregation, and surface area—from the same mineral (e.g., crocidolite and nonasbestiform riebeckite) are needed for a variety of mineral types. Ultimately, management of risk needs to be based on an understanding of the estimated quantitative risk from exposures to various EMPs and on an understanding of the concentrations and particle characteristics of the various EMPs to which humans are or could be exposed and the exposures in occupational and other settings.

An important need is to identify and develop analytical methods that can be used or modified to quantify occupational exposures to EMPs of toxicological importance and that are capable of differentiating EMPs on the basis of particle characteristics demonstrated to be important in causing disease. The current PCM method is inadequate for assessing exposures to fibers in mixed-dust environments, and it lacks the capability to measure all the important physical and chemical parameters of EMPs thought to be associated with toxicity. The feasibility of using EM methods will need to be addressed if they are to replace the PCM method. Until these issues are addressed, improvements in PCM methodologies should be pursued. In epidemiological and toxicological research, EM methods will need to be used to carefully characterize the exposure materials. Also, the results of toxicological and epidemiological studies may identify additional particle characteristics that warrant further evaluation to determine whether they can be incorporated into sampling and analytical methods used to assess worker exposures.

Section 3 of this *Roadmap* presents a framework for proposed research intended to address these scientific issues and inform future public health policies and practices.

3 Framework for Research

3.1 Strategic Research Goals and Objectives

Strategic goals and objectives for a multidisciplinary research program on mineral fibers and other EMPs are identified below. Shown in brackets following each goal and objective is the number of the section of this *Roadmap* in which the goal or objective is subsequently discussed.

I. **Develop a broader understanding of the important determinants of toxicity for asbestos fibers and other EMPs [3.2].**

- Conduct *in vitro* studies to ascertain what physical, chemical, surface properties, and other particle characteristics influence the toxicity of asbestos fibers and other EMPs [3.2.1]; and
- Conduct animal studies to ascertain what physical and chemical properties, surface properties, and other particle characteristics influence the toxicity of asbestos fibers and other EMPs [3.2.2].

II. **Develop information and knowledge on occupational exposures to asbestos fibers and other EMPs and related health outcomes [3.3].**

- Assess available occupational exposure information relating to various types of asbestos fibers and other EMPs [3.3.1];
- Collect and analyze available information on health outcomes associated with exposures to various types of asbestos fibers and other EMPs [3.3.2];
- Conduct selective epidemiologic studies of workers exposed to various types of asbestos fibers and other EMPs [3.3.3]; and
- Improve clinical tools and practices for screening, diagnosis, treatment, and secondary prevention of diseases caused by asbestos fibers and other EMPs [3.3.4].

III. **Develop improved sampling and analytical methods for asbestos fibers and other EMPs [3.4].**

- Reduce inter-operator and inter-laboratory variability of the current analytical methods used for asbestos fibers [3.4.1];
- Develop analytical methods with improved sensitivity to visualize thinner EMPs to ensure a more complete evaluation of airborne exposures [3.4.2];
- Develop a practical analytical method for air samples to differentiate between exposures to asbestiform fibers from the asbestos minerals and exposures to EMPs from their nonasbestiform analogs [3.4.3];
- Develop analytical methods to assess durability of EMPs [3.4.4]; and
- Develop and validate size-selective sampling methods for EMPs [3.4.5].

3.2 An Approach to Conducting Interdisciplinary Research

Within each of the goals and objectives laid out in this framework, a more detailed research program will have to be developed. Research to support these three goals must be planned and conducted by means of an interdisciplinary approach, incorporating toxicological, epidemiological, exposure assessment, medical, analytical, and mineralogical methods. The research must also be integrated to optimize resources, facilitate the simultaneous collection of data, and ensure, to the extent feasible, that the research builds toward a resolution of the key issues. An aim of the research is to acquire a level of mechanistic understanding that can provide the basis for developing biologically-based models for extrapolating results of animal inhalation and other types of *in vivo* studies to exposure conditions typically encountered in the workplace. The information gained from such research can then be used by regulatory agencies and occupational health professionals to implement appropriate exposure limits and risk management programs for monitoring worker exposure and health. Much of this research may be accomplished by NIOSH, other federal agencies, or other stakeholders. Any individual research project undertaken should be designed to ensure that the results can be interpreted and applied within the context of other studies in the overall program and lead to outcomes useful for decision-making and policy-setting.

3.3 National Reference Repository of Minerals and Information System

To support the needed research, a national reference repository of samples of asbestos and related minerals will be required, and a database of relevant information should be developed. Minerals vary in composition and morphology by location and origin, and differences within the same mineral type can be significant. Currently, no national repository exists to retain, document, and distribute samples of asbestiform and nonasbestiform reference minerals for research and testing. These reference samples should be well-characterized research-grade materials that are made available to the research community for testing and standardization. This will allow minerals with matching properties (e.g., morphology, dimension) to be chosen for study. To accomplish this research, exhaustive characterization of the samples, including contaminants, is necessary. Detailed characterizations of minerals should include their ability to affect biological activities and properties such as (1) purity, (2) particle morphology (range of dimensions and sizes), (3) surface area, (4) surface chemistry, and (5) surface reactivity. The particle characteristics identified by Hochella [1993] should be considered for particle characterization. The use of these samples in research would facilitate meaningful comparisons and reduce uncertainties in the interpretation of results between and among studies.

Care must be taken to ensure that a sufficient amount of the studied material is available, not only for current studies, but also as reference material for possible future studies. The repository should accept and store samples from researchers conducting EMP studies so that further testing and characterizations can be performed. The information developed from all of these efforts should be entered into a database, which can serve as a tool for selection of minerals for testing and validation of toxicological tests, as well as to assist in identification of worker populations for possible epidemiological studies.

The development of a comprehensive, publicly available information system incorporating all studies of the toxicity, exposures, and health effects of asbestos and related minerals could help enhance the development of the research programs, avoid duplication of effort, and enhance interpretation of the data generated. The system should include all pertinent information about methods, doses or exposures, minerals, particle characteristics, and other topics.

3.4 Develop a Broader Understanding of the Important Determinants of Toxicity for Asbestos Fibers and Other EMPs

To address this objective, one of the first steps will be to identify the range of minerals and mineral habits needed to systematically address the mineral characteristics that may determine particle toxicity. Care must be taken to ensure that mineralogical issues in a study are adequately addressed. Information on both crystalline lattice structure and composition is needed to define a mineral species, because information on either alone is insufficient to describe the properties of a mineral. For example, nonasbestiform riebeckite and asbestiform riebeckite (crocidolite) share the same elemental composition but have different expressions of their crystalline lattices. EMPs from nonasbestiform riebeckite are not flexible. Crocidolite fibers generally have chain-width defects, which explain their flexibility. These chain-width defects also affect diffusion of cations and dissolution properties, both of which can explain greater release of iron into surrounding fluid by crocidolite than by nonasbestiform riebeckite [Guthrie 1997].

In addition to elemental content and crystalline habit, the particle characteristics identified by Hochella [1993] should be considered for particle characterization. For example, the current paradigm for fiber pathogenicity does not discriminate between different compositions of biopersistent fibers, except insofar as composition determines biopersistence. Two biopersistent fiber types, erionite [Wagner et al. 1985] and silicon carbide [Davis et al. 1996], show a special proclivity to cause mesothelioma, for reasons that are not easily explained by the current paradigm because their biopersistence and distributions of fiber lengths are not substantially different from those of the amphiboles. The biochemical basis of the enhanced pathogenicity of these two fiber types has not been elucidated. This suggests that some fiber types may possess surface or chemical reactivity that imparts added pathogenicity over and above what would be anticipated for long biopersistent fibers. Because of the many variations in elemental composition, crystalline structure, and other characteristics of these minerals, it will be impossible to study all variants. Therefore, a strategy will need to be developed for selecting minerals for testing. This strategy should include consideration of occupationally relevant minerals and habits in order to complement available information, availability of appropriate and well-characterized specimens for testing, and practical relevance of the results to be achieved through testing.

EPA's Office of Pollution Prevention and Toxics, NIEHS, NIOSH, and OSHA assembled an expert panel in the 1990s to consider major issues in the use of animal models for chronic inhalation toxicity and carcinogenicity testing of thoracic-size elongate particles. Issues considered included the design of animal tests of chronic inhalation exposure to EMPs; preliminary studies to guide them; parallel mechanistic studies to help interpret study results and to extrapolate findings with regard to potential health effects in humans; and available screening tests for identifying and assigning a priority for chronic inhalation study. There was general agreement that (1) chronic inhalation studies of EMPs in the rat are the most appropriate tests for predicting inhalation hazard and risk of EMPs to humans; (2) no single assay or battery of short-term assays could predict the outcome of a chronic inhalation bioassay for carcinogenicity; and (3) several short-term *in vitro* and *in vivo* studies may be useful to assess the relative potential of various EMPs to cause lung toxicity or carcinogenicity [Vu et al. 1996].

Such short-term assays and strategies were considered by an expert working group assembled by the International Life Sciences Institute's Risk Science Institute to arrive at a consensus on current assays useful for screening EMPs for potential toxicity and carcinogenicity [ILSI 2005]. Dose, dimension, durability, and possible surface reactivities were identified as critical parameters for study, although it was noted that no single physicochemical property or mechanism can now be used to predict carcinogenicity of all EMPs. The strategy for short-term (i.e., 3 months or less) testing in animal models included sample preparation and characterization (composition, crystallinity, habit, size distribution); testing for biopersistence *in vivo* with use of a standard protocol such as that of the European Union [European Commission 1999]; and a subchronic inhalation or instillation challenge of the rat, with evaluation of lung weight and fiber burden, bronchoalveolar lavage profile, cell proliferation, fibrosis, and histopathology. Additionally, other nonroutine analyses for particle surface area and surface reactivities and short-term *in vitro* cellular toxicological assays might be evaluated. The use of *in vitro* tests should be tempered by the observations that standard protocols fail to distinguish relative pathogenic potentials of even nonelongate silicates (such as quartz versus clay dusts) and that treatment of particle surfaces (such as modeling their conditioning upon deposition on the lipoprotein-rich aqueous hypophase surface of the deep lung) can greatly affect their expression of toxicities [ATSDR 2003].

EMPs encountered in any particular work environment are frequently heterogeneous, which limits the ability of epidemiological and other types of health assessment studies to evaluate the influence on toxicity of EMP dimensions (length and width), chemical composition, biopersistence, and other characteristics. Toxicological testing is needed to address some of the fundamental questions about EMP toxicity that cannot be determined through epidemiology or other types of health assessment studies. Irrespective of study type or design, the full characterization of all particulate material in a test sample is an essential step in understanding the mechanisms of EMP toxicity. The determination of EMP dimensions is important, and they are best expressed as bivariate size distributions (i.e., width and length). Such determinations should be made by using both relatively simple procedures (optical microscopy) and highly specialized techniques (e.g., TEM or SEM with EDS), because size-specific fractions of EMP exposures have both biological and regulatory significance.

The chemical composition (e.g., intrinsic chemical constituents and surface chemistry) of asbestos fibers and other EMPs has been shown to have a direct effect on their ability to persist in the lung and to interact with surrounding tissue to cause DNA damage. For example, ferric and ferrous cations are major components of the crystalline structure of amphibole asbestos fibers; iron may also be present as a surface impurity on chrysotile asbestos fibers and other EMPs. The availability of iron at the surface of asbestos fibers and other EMPs has been shown to be a critical parameter in catalyzing the generation of ROS that may indirectly cause genetic damage [Kane 1996]. Also, attempted clearance of long asbestos fibers from the lung causes frustrated phagocytosis, which stimulates the release of ROS [Mossman and Marsh 1989]. Individual adaptive responses to oxidant stress and the body's ability to repair damaged DNA are dependent on multiple exogenous and endogenous factors, but few experiments have been attempted to evaluate these variables in animal or human model systems. Kane [1996] has suggested that the mechanisms responsible for the genotoxic effects of asbestos fibers are due to indirect DNA damage mediated by free radicals and direct physical interference with the mitotic apparatus by the fibers themselves. Research to address the following questions would assist in validating these proposed mechanisms:

- Are *in vitro* genotoxicity assays relevant to carcinogenesis of asbestos fibers and other EMPs?
- Are *in vitro* doses relevant for *in vivo* exposures?
- Can genotoxic effects of asbestos fibers and other EMPs be assessed *in vivo*?

Macrophages are the initial target cells of EMPs and other particulates deposited in the lungs or pleural and peritoneal spaces. Phagocytosis of asbestos fibers has been shown to be accompanied by the activation of macrophages, which results in the generation of ROS as well as a variety of chemical mediators and cytokines [Kane 1996]. These mediators amplify the local inflammatory reaction. Persistence of asbestos fibers in the lung interstitium or in the subpleural connective tissue may lead to a sustained chronic inflammatory reaction accompanied by fibrosis [Oberdorster 1994]. The unregulated or persistent release of these inflammatory mediators may lead to tissue injury, scarring by fibrosis, and proliferation of epithelial and mesenchymal cells. This response appears to occur independently of fiber length. In the lungs and pleural linings, chronic inflammation and fibrosis are common reactions following exposure to asbestos fibers, but research is needed to understand the effects of fiber dimension, loading, and surface reactivity on the development of disease endpoints such as inflammation, fibrosis, and cancer.

It has been suggested that asbestos fibers and other EMPs may contribute to carcinogenesis by multiple mechanisms and that EMPs may act at multiple stages in neoplastic development, depending on their physicochemical composition, surface reactivity, and biopersistence in the lung [Barrett 1994]. Animal inhalation studies are needed to investigate the biopersistence and toxicity of asbestos fibers and other EMPs representing a range of chemical compositions and morphological characteristics (including crystalline habits) and representing a range of discrete lengths and widths. An additional factor that should be considered and evaluated is the influence of concurrent exposure to other particles and contaminants on the biopersistence and toxicity of EMPs. In a recently reported short-term (5-day) animal inhalation study to evaluate the biopersistence of chrysotile fibers with and without concurrent exposure to joint compound particles (1–4 μm MMAD), the clearance half-time of all fiber sizes was approximately an order of magnitude less for the group exposed to chrysotile and joint-compound particles [Bernstein et al. 2008]. Histopathological examination indicated that the combination of chrysotile and fine particles accelerated the recruitment of alveolar macrophages, resulting in a 10-fold decrease in the number of fibers remaining in the lung. Although no mention was made of any pathological changes in the lungs of the chrysotile/particulate-exposed group, other studies have shown that the recruitment of macrophages then increases the production and recruitment of polymorphonuclear leukocytes, which themselves can generate ROS [Driscoll et al. 2002; Donaldson and Tran 2002].

Much research has been focused on lung cancer and mesothelioma. Even if it is determined that EMPs from some minerals have low potency for causing cancer, additional studies may be needed to investigate their potential for causing inflammation, fibrosis, and other nonmalignant respiratory effects. Also, the relationship between EMP dimension and fibrosis should be more fully investigated. The results of such research may allow currently used standard exposure indices to be modified by specifying different dimensional criteria (lengths and widths) relevant to each of the disease outcomes associated with EMP exposures, and by determining whether biopersistence should be included as an additional criterion. However, this research may be dependent on the development of new technology that can generate mineral fibers and other EMPs of specific dimensions in sufficient quantities to conduct animal inhalation experiments. Consequently, the development of revised exposure indices based on EMP dimension may not be possible in the short term.

The research strategy described above should conform with the general strategies and tactics that have been recommended by several expert panels for clarifying the risks and causes of asbestos exposure–associated diseases and with the current effort of the U.S. Federal Government Interagency Asbestos Working Group (IAWG), involving participation of the EPA, USGS, NIOSH, ATSDR, CPSC, OSHA, MSHA, and NIEHS/NTP, to identify federal research needs and possible actions regarding asbestos fibers and other durable EMPs of public health concern [Vu et al. 1996; ILSI 2005; Schins 2002; Greim 2004; Mossman et al. 2007].

An ILSI Risk Science Institute Working Group supported by EPA published a tiered testing strategy for fibrous particles in 2005 [ILSI 2005]. Consideration should be given to the following slight modification of this published scheme.

Noteworthy in the findings of the ILSI Working Group report is the inadequacy of *in vitro* test models to predict the *in vivo* toxicity of EMPs. Indeed, many man-made mineral fibers are positive in cell test systems but do not cause fibrosis or cancer in chronic animal models. The *in vitro* test systems lack predictive ability because they do not incorporate biopersistence. For this reason, *in vitro* tests, other than assays for durability, are not included in the tiered testing strategy given below.

Step 1. Preparation and characterization of test EMPs

This is the initial, required step for any toxicological evaluation. It should include

- comprehensive chemical and mineralogical characterization, including crystallinity and EMP habit.
- size distribution of the EMPs found in the workplace (total particulate sample), as well as dimensional characteristics of size-selected fraction(s) to be used for hazard evaluation. A limiting step for detailed toxicological evaluation is the availability of sufficient quantities of size-selected EMPs of known chemistry and mineralogy.

Step 2. Assessment of *in vitro* durability

Evidence indicates that highly soluble fibrous particles do not exhibit fibrotic or carcinogenic potential in animal studies. Rate of dissolution in simulated body fluids should be measured with a dynamic flow-through system, as outlined by Potter et al. [2000]. In brief, EMPs are exposed by continuous flow to a modified Gamble's solution, and fiber diameter is monitored optically over time. Biopersistence would be an indication of concern and would indicate the need for further testing of the pathogenic potential of the EMP. This step is optional, as one could move directly to Step 3.

Step 3. Short-term *in vivo* biopersistence test

Biopersistence of fibers longer than 20 μm has been found to be an excellent predictor of collagen deposition in chronic inhalation studies [Bernstein et al. 2001]. Two alternative methods are accepted by the European Commission [1997]: intratracheal instillation and 5-day inhalation by rats. It is recommended that fiber burden be measured at time points up to 3 months post-exposure. Biopersistence would be an indication of concern and would indicate the need for further testing of the pathogenic potential of the EMP.

Step 4. Subchronic inhalation study

Parameters that should be measured in such an inhalation study are noted by EPA [2001]. The test should involve inhalation exposure for 3 months and should evaluate pulmonary responses over 6 months post-exposure. Responses to be measured should include biopersistence, persistent inflammation, cell proliferation (bromodeoxyuradine [BrdU] assay), fibrosis, epithelial cell hyperplasia, lung weight, and fiber burden. Biopersistence and persistent inflammation are notable markers of concern. If the subchronic study is positive, a long-term inhalation study is necessary to conduct a full-risk assessment.

Step 5. Long-term inhalation study

The test would include a 2-year inhalation study in rats, with lifelong follow-up. Fibrosis, lung tumors, and mesothelioma should be measured according to EPA guidelines [EPA 2001] for long-term inhalation studies of fibers. Obtainment of lung burden, dose-related response, and time-course data would enable risk assessment.

Implicit in any new or revised occupational health policy for EMPs would be the need to conduct appropriate assessments of risk. Risk assessments for lung cancer, mesothelioma, and asbestosis have been conducted on worker populations exposed to various asbestos minerals. These risks have been qualitatively confirmed in animals, but no adequate quantitative multidose inhalation studies with asbestos have been conducted in rodents that would permit direct comparisons to lung cancer and mesothelioma risks determined from exposed human populations. Given the availability of risk estimates for lung cancer in asbestos-exposed humans, chronic studies with rats exposed to asbestos (e.g., chrysotile) fibers would provide an assessment of the rat as a valid "predictor" of human lung cancer risks associated with exposure to asbestos fibers and other EMPs.

3.4.1 Conduct *In Vitro* Studies to Ascertain the Physical and Chemical Properties that Influence the Toxicity of Asbestos Fibers and Other EMPs

Although *in vitro* studies may not be appropriate for toxicology screening tests of EMPs, they can help clarify the mechanisms by which some EMPs induce cancer, mesothelioma, or fibrosis, as well as the properties of EMPs and conditions of exposure that determine pathogenicity. *In vitro* studies allow specific biological and mechanistic pathways to be isolated and tested under controlled conditions that are not feasible in animal studies. *In vitro* studies can yield data rapidly and provide important insights and confirmations of mechanisms, which can be confirmed with specifically designed *in vivo* studies.

With the exception of *in vitro* genotoxicity testing of asbestos fibers, little testing information is available on the potential genotoxicity of EMPs. In contrast to standard genotoxicity testing of soluble substances, the results from testing EMPs can be influenced by dimension, surface properties, and biopersistence. The mechanisms of asbestos-induced genotoxicity are not clear, but direct interaction with the genetic material and indirect effects via production of ROS have been proposed. A combination of the micronucleus test and the comet assay with continuous treatment (without exogenous metabolic activation) has been reported to detect genotoxic activity of asbestos fibers [Speit 2002]. However, further research is needed to determine whether this approach is applicable for genotoxicity testing of other EMPs. Before conducting such studies, the following EMP interactions should be addressed:

- initial lesions evoking cell damage or response (e.g., direct or indirect cytotoxic or genotoxic events or induction of toxic reactive intermediate materials);
- subsequent multistage cellular response (e.g., intracellular signaling through a kinase cascade to nuclear transcription of factors for apoptosis, cell transformation, and cell or cell system proliferation or remodeling and initiation or promotion of neoplasia or fibrosis); and
- critical time-course events in those processes (e.g., cell-cycle-dependent EMP interactions or EMP durability under different phagocytic conditions).

Capabilities for conducting these studies improved during the 2000s, including

- advancement in analytical methods for physicochemical characterization of EMP properties (e.g., for resolving small dimensions and nanoscale surface properties); and

- ability to prepare EMP samples that are "monochromatic" in size or surface properties in quantities sufficient for well-controlled *in vitro* assays.

Identification of the initiating EMP-cell interactions calls for research on the mechanisms of

- cell-free generation of toxic ROS by EMPs or EMP-induced cellular generation of toxic ROS; and
- direct membranolytic, cytotoxic, or genotoxic activities of the EMP surface in contact with cellular membranes or genetic material.

These investigations will require attention to the following:

- effects of EMP surface composition (e.g., surface-borne iron species);
- effects of normal physiological conditioning of respired particles (e.g., *in vitro* modeling of *in vivo* initial conditioning of EMP surfaces by pulmonary surfactant);
- nonphysiological conditioning of EMP under *in vitro* test conditions (e.g., by components of nutrient medium);
- cell type (e.g., phagocytic inflammatory cell or phagocytic versus nonphagocytic target cell); and
- EMP dimensions in relation to cell size (e.g., as a factor distinguishing total phagocytosis and partial frustrated phagocytosis).

Cell generation of ROS is seen generally in phagocytic uptake of elongate or nonelongate particles (e.g., as a respiratory burst). In normal phagocytosis, there is a maturation of the phagosomal membrane, with progression to a phagolysosomal structure for attempted lysosomal digestion. Anomalous behavior of this system may occur in frustrated phagocytosis of long EMPs. The hypothesis of frustrated phagocytosis suggests that EMPs that are too long to permit full invagination may stimulate cells to generate ROS or anomalously release lytic factors into the extracellular annulus rather than into a closed intracellular phagosome.

EMP surfaces may be tested for direct membranolytic or cytotoxic activities that are dependent on surface composition or structure. As a guide, membranolytic or cytotoxic activities of nonelongate particulate silicates are dependent on surface properties. Nonelongate particulate silicates also provide an example of failure of *in vitro* cytotoxicity to relate with pathogenicity (e.g., respirable particles of quartz or kaolin clay significantly differ in disease risk for fibrosis, but are comparably cytotoxic *in vitro* unless they are preconditioned with pulmonary surfactants and then subjected to phagolysosomal digestion). *In vitro* studies of direct versus indirect induction of genotoxic activities may consider factors affecting the bioavailability of the nuclear genetic material (e.g., the state of phagocytic activity of the cell or the stages in the cell cycle with collapse of the nuclear membrane in mitosis). These again suggest care in the preparation of EMPs and the manner of challenge with EMPs employed in *in vitro* experiments.

The two modes of primary damage, a release of reactive toxic agents induced by long particulates or a surface-based membranolytic or genotoxic mechanism, may be involved singly or jointly in primary cell responses to EMPs. These may be investigated by comparing the effects of different types of EMPs (e.g., relative potencies of erionite fibers and amphibole asbestos fibers in *in vitro* cell transformation studies are different than their potencies in *in vivo* induction of mesothelioma).

In the second phase of cellular response to EMPs, the central dogma of intracellular response is being intensively researched. The initial extracellular primary damage induces intracellular signaling (e.g., by MAPK), which causes a cascade of kinase activities that stimulate selective nuclear transcription of mRNAs, leading to production of TNF-α or other cytokines for extracellular signaling of target cells. Those other cytokines may induce cell proliferation toward cancer or collagen synthesis toward fibrosis. Further definition of signaling mechanisms and analyses of their induction by different primary EMP-cellular interactions may better define the ultimate role of EMP properties in the overall process. That research, again, may be facilitated by using different specific types of EMPs, each with relatively homogeneous morphology and surface properties.

Although full investigation of biopersistence of EMPs may require long-term animal model studies, *in vitro* systems coupled with advanced surface analytical tools (e.g., field emission SEM–energy dispersive X-ray spectroscopy or scanning Auger spectroscopy) may help guide *in vivo* studies. This could be done by detailing specific surface properties of EMPs and their modifications under cell-free or *in vitro* conditions representing the local pH and reactive species at the EMP surface under conditions of extracellular, intra-phagolysosomal, or frustrated annular phagocytic environments.

3.4.2 Conduct Animal Studies to Ascertain the Physical and Chemical Properties that Influence the Toxicity of Asbestos Fibers and Other EMPs

A multispecies testing approach has been recommended for short-term assays [ILSI 2005] and chronic inhalation studies [EPA 2000] that would provide solid scientific evidence on which to base human risk assessments for a variety of EMPs. To date, the most substantial base of human health data for estimating lung cancer risk is for workers exposed to fibers from different varieties of asbestos minerals.

There are similarities between animal species and humans in deposition, clearance, and retention of inhaled particles in the lungs. For example, peak alveolar deposition fraction is greatest for particles at an AED of about 2 μm in humans and about 1 μm for rodents. However, some interspecies differences have been identified. Variations in ventilation, airway anatomy, and airway cell morphology and distribution account for quantitative differences in deposition pattern and rate of clearance among species. In addition, the efficacy of pulmonary macrophage function differs among species. All these differences could affect particle clearance and retention. It has been suggested that the following species differences should be considered in the design of experimental animal inhalation studies of elongate particles [Dai and Yu 1988; Warheit et al. 1988; Warheit 1989].

- Due to differences in airway structure, airway size, and ventilation parameters, a greater fraction with large AED particles are deposited in humans than in rodents.
- Alveolar deposition fraction in humans varies with workload. An increase in the workload reduces the deposition fraction in the alveolar region because more of the inhaled particulate is deposited in the extra-thoracic and bronchial regions.
- Mouth breathing by humans results in a greater upper bronchial deposition and enhanced particle penetration to the peripheral lung.

- For rats and hamsters, alveolar deposition becomes practically zero when particle AED exceeds 3.0 μm and aspect ratio exceeds 10. In contrast, considerable alveolar deposition is found for humans breathing at rest, even for EPs with AEDs approaching 5 μm and aspect ratio exceeding 10.

- Rodents have smaller-diameter airways than humans, a characteristic which increases the chance for particle deposition via contact with airway surfaces.

- Turbulent air flow, which enhances particle deposition via impaction, is common in human airways but rare in rodent airways.

- Variations in airway branching patterns may account for significant differences in deposition between humans and rodents. Human airways are characterized by symmetrical branching, wherein each bifurcation is located near the centerline of the parent airway. This symmetry favors deposition "hotspots" on carinal ridges at the bifurcations due to disrupted airstreams and local turbulence. Rodent airways are characterized by asymmetric branching, which results in a more diffuse deposition pattern because the bulk flow of inspired air follows the major airways with little change in velocity or direction.

- Alveolar clearance is slower in humans than in rats. Human dosimetry models predict that, at nonoverloading exposure concentrations, a greater proportion of particles deposited in the alveolar region will be interstitialized and sequestered in humans than in rats.

An important consideration in the conduct and interpretation of animal studies is the selection of well characterized (with respect to chemical and physical parameters) and appropriately sized EMPs that take into account differences in deposition and clearance characteristics between rodents and humans. EMPs that are capable of being deposited in the bronchoalveolar region of humans cannot be completely evaluated in animal inhalation studies because the maximum thoracic size for particles in rodents is approximately 2 μm AED, which is less than the maximum thoracic size of about 3 μm AED for humans [Timbrell 1982; Su and Cheng 2005].

3.4.2.1 Short-Term Animal Studies

There are several advantages to conducting short-term animal studies in rats; one is that the information gained (e.g., regarding overload and maximum tolerated dose [MTD]) can be used to more effectively design chronic inhalation studies [ILSI 2005]. The objectives of these studies would be to

- Evaluate EMP deposition, translocation, and clearance mechanisms;

- Compare the biopersistence of EMPs retained in the lung with results from *in vitro* durability assays;

- Compare *in vivo* pulmonary responses to *in vitro* bioactivity for EMPs of different dimensions; and

- Compare cancer and noncancer toxicities for EMPs from asbestiform and nonasbestiform amphibole mineral varieties of varying shapes, as well as within narrow ranges of length and width.

More fundamental studies should also be performed to

- Identify biomarkers or tracer/imaging methods that could be used to predict or monitor active pulmonary inflammation, pulmonary fibrosis, and malignant transformation;

- Investigate mechanisms of EMP-induced pulmonary disease; and
- Determine whether cell proliferation in the lungs (terminal bronchioles and alveolar ducts) can be a predictive measure of pathogenicity following brief inhalation exposure and use of the BrdU assay [Cullen et al. 1997].

Exposure protocols for tracheal inhalation or instillation in an animal model for short-term *in vivo* studies using field-collected or laboratory-generated EMPs should address possible adulteration of EMP morphology (e.g., anomalous agglomeration of particles). This might be addressed in part by preconditioning EMPs in a delivery vehicle containing representative components of pulmonary hypophase fluids. Exposure protocols using pharyngeal aspiration as a delivery system should be considered, given the observations in studies with single-walled carbon nanotubes that such a delivery system closely mimics animal inhalation studies [Shvedova et al. 2005, 2008].

Studies evaluating the roles of biopersistence and dimension in the development of noncancer and cancer endpoints from exposure to EMPs are also needed. These studies should attempt to elucidate the physicochemical parameters that might affect bio-durability of EMPs of specific dimensions. Although short-term animal inhalation studies would be informative, companion *in vitro* assays should also be conducted to assess their validity for screening EMPs.

3.4.2.2 Long-Term Animal Studies

Chronic animal inhalation studies are required to address the impacts of dimension, morphology, chemistry, and biopersistence on critical disease endpoints of cancer induction and nonmalignant respiratory disease. The EPA's proposed testing guidelines should be considered as the criteria for establishing the testing parameters for chronic studies [EPA 2001].

To date, chronic inhalation studies have been conducted with different animal species and different types of EPs. However, it remains uncertain which species of animal(s) best predict(s) the risk of respiratory disease(s) for workers exposed to different EPs. Chronic inhalation studies should be initiated to establish exposure/dose-response relationships for at least two animal species. The rat has historically been the animal of choice for chronic inhalation studies with EPs, but the low incidence of lung tumors and mesotheliomas in rats exposed to asbestos fibers suggests that rats may be less sensitive than humans. Therefore, any future consideration for conducting long-term animal inhalation studies should address the need for using a multispecies testing approach to help provide solid scientific evidence on which to base human risk assessments for a variety of EMPs of different durabilities and dimensions. For example, some recent studies suggest that the hamster may be a more sensitive model for mesothelioma than the rat. Validation of appropriate animal models could reduce the resources needed to perform long-term experimental studies on other EMP types [EPA 2001].

Multidose animal inhalation studies with asbestos (probably a carefully selected and well-characterized chrysotile, because most of the estimates of human risk have been established from epidemiological studies of chrysotile-exposed workers) are needed to provide an improved basis for comparing the potential cancer and noncancer risks associated with other types of EMPs and various types of synthetic EPs. The asbestos fibers administered in these animal studies should be comparable in dimension to those fibers found in the occupational environment. The results from these studies with asbestos (e.g., chrysotile) would provide a "gold standard" that could be used to validate the utility of long-term inhalation studies (in rats or other species) for predicting human risks of exposure to various types of EMPs.

3.4.3 Evaluate Toxicological Mechanisms to Develop Early Biomarkers of Human Health Effects

The following scheme using acellular and cellular tests can be conducted to develop a mechanistic understanding of fiber toxicity and to support the development of *in vivo* biomarkers of effect in humans. These studies must use well-characterized EMP samples as described in the tiered testing strategy presented in Section 3.4. The use of size-selected fractions of EMPs could provide information needed to understand the relationships between dimension and bioactivity.

Acellular assays could include measurement of the generation of ROS employing electron spin resonance (ESR) or oxidant sensitive fluorescent dyes. Evaluation of the mobilization of metal ions from EMPs could indicate cytotoxic potential.

The *in vitro* cellular tests could include the following:
- generation of reactive species measured by ESR or fluorescent dyes;
- generation of inflammatory, fibrogenic, and proliferative mediators, such as TNF-alpha, IL-1, and TGF;
- DNA damage by comet assay;
- effects on cell growth regulation by measuring cell proliferation;
- effects on mitosis and aneuploidy, determined by confocal fluorescent microscopy; and
- signal transduction pathways, such as MAPK, and phosphoinositide-3 (PI3) kinase pathways.

In vivo tests would measure markers of inflammation (e.g., BAL neutrophils, inflammatory cytokines and chemokines), fibrosis (collagen, hydroxyproline), and proliferation (BrdU assay, hyperplasia) that precede pathology. Knockout mice or pathway inhibitors in rats may be used to confirm mechanistic pathways identified *in vitro* and develop biomarkers for disease initiation and progression. Potential biomarkers identified in *in vitro* and *in vivo* studies would be evaluated in human populations with known exposure to EMPs, and the type and extent of the relationships between the marker and clinical signs of disease could be determined.

3.5 Develop Information and Knowledge on Occupational Exposures to Asbestos Fibers and Other EMPs and Related Health Outcomes

Many studies have been published concerning occupational exposures to asbestos fibers and associated health effects. These studies have formed a knowledge base that has supported increased regulation of occupational asbestos exposures and substantial reductions in asbestos use and asbestos exposures in the United States over the past several decades. But, as this *Roadmap* makes clear, much less is known about other types of mineral fibers and EMPs in terms of occupational exposures and potential health effects.

Research is needed to produce information on

- current estimates and, where possible, future projections of numbers of U.S. workers exposed to asbestos fibers;
- levels of current exposures and nature of the exposures (e.g., continuous, short-term, or intermittent); and
- the nature of any concomitant dust exposures.

Similar research is needed to produce analogous information about occupational exposures to other EMPs. Research is needed to assess and quantify potential human health risks associated with occupational exposures to other EMPs, as well as to better understand and quantify the epidemiology of asbestos-related diseases by using more refined indices of exposure. Research is also needed to produce improved methods and clinical guidance for screening, diagnosis, secondary prevention, and treatment of diseases caused by asbestos fibers and other hazardous EMPs.

3.5.1 Assess Available Information on Occupational Exposures to Asbestos Fibers and Other EMPs

A fully informed strategy for prioritizing research on EMPs should be based on preliminary systematic collection and evaluation of available information on (1) industries/occupations/job tasks/processes with exposure to various types of asbestos fibers and other EMPs; (2) numbers of workers exposed; (3) characteristics and levels of exposures; and (4) associated concomitant particulate exposures. Such information could enable estimations of

- the overall distribution and levels of occupational exposures and an estimate of the total number of workers exposed to EMPs currently, in the past, and projected in the future; and
- the specific distributions and levels of exposures to each particular type of EMP, as well as numbers of workers exposed to each type of EMP currently, in the past, and projected in the future.

To complement readily available information disseminated by the USGS on annual domestic production and importation of raw asbestos, data on the annual amounts of asbestos imported into the United States in the form of any asbestos-containing products should be aggregated and made easily accessible.

Additional efforts should be made to collect, review, and summarize available occupational exposure information and to collect and analyze representative air samples relating to various types of EMPs. For example, systematic compilation of exposure data collected by OSHA, MSHA, NIOSH, state agencies, and private industry could contribute to an improved understanding of current occupational exposures to EMPs, particularly if there are opportunities to (re)analyze collected samples with use of enhanced analytical methods to better characterize the exposures (see Section 3.6). To help limit potential impact of sampling bias that may be inherent in the available EMP exposure data, these initial efforts should be supplemented with efforts to systematically identify, sample, and characterize EMP exposures throughout U.S. industry. These exposure assessments should include workplaces in which a fraction of the dust comprises EMPs (i.e., mixed-dust environments), and occupational environments in which EMPs may not meet the current regulatory criteria to be counted (i.e., "short" fibers). With appropriate planning and resources, such efforts could be designed and implemented as ongoing surveillance of occupational exposures to EMPs, with periodic summary reporting of findings. Representative EMP exposure data could help identify worker populations or particular types of EMPs warranting further study (i.e., more in-depth exposure assessment; medical surveillance; epidemiologic studies of particular types of EMPs, processes, job tasks, occupations, or industries; and toxicity studies of particular EMPs). Occupational exposure data should be collected and stored in a comprehensive database. Information similar to that described by Marchant et al. [2002] should be incorporated into the database to support these efforts. This could be accomplished in parallel with efforts to develop an occupational exposure database for nanotechnology [Miller et al. 2007] or efforts to develop a national occupational exposure database [Middendorf et al. 2007].

3.5.2 Collect and Analyze Available Information on Health Outcomes Associated with Exposures to Asbestos Fibers and Other EMPs

The body of knowledge concerning human health effects from exposure to EMPs consists primarily of findings from epidemiological studies of workers exposed to asbestos fibers. There is general agreement that workers exposed to fibers from any asbestiform amphibole mineral would be at risk of serious adverse health outcomes of the type caused by exposure to fibers from the six commercially exploited asbestos minerals. NIOSH commented on the most recently proposed MSHA rule on asbestos (subsequently promulgated as a final rule), stating that "NIOSH remains concerned that the regulatory definition of asbestos should include asbestiform mineral fibers such as winchite and richterite, which were of major importance as contaminants in the Libby, MT, vermiculite" [NIOSH 2005]. To ensure a clear science base that might support a formal recommendation for control of occupational exposures to all asbestiform amphibole fibers, it would be reasonable to thoroughly review, assess, and summarize the available information on asbestiform amphiboles such as winchite, richterite, and fluoro-edenite that have not been commercially exploited.

It will also be important to quantitatively determine the health risks posed by EMPs from nonasbestiform amphiboles and to compare them to the risks posed by fibers from asbestiform amphiboles. If nonasbestiform amphibole EMPs are, in fact, associated with some risk, a quantitative risk assessment would be needed to understand the risks relative to those associated with exposures to asbestos fibers. If new epidemiological and other evidence is sufficient to support such a risk estimate, that could lead to the development of risk management policies for nonasbestiform amphibole EMPs distinct from risk management policies for asbestos fibers. Risk management policies that differ for asbestiform and nonasbestiform amphiboles would motivate the development and routine use of new analytical methods that differentiate asbestiform from nonasbestiform particles.

In addition to collecting and analyzing information on health outcomes associated with exposures to asbestiform EMPs and EMPs from nonasbestiform amphiboles, similar information related to other EMPs (e.g., erionite, wollastonite, attapulgite, and fibrous talc) should be collected for systematic analysis. Additional relevant information may be gleaned from epidemiological studies conducted on some SVFs (e.g., glass and mineral wool fibers, ceramic fibers).

Surveillance and epidemiological studies generally have been circumscribed by the long latency periods that characterize manifestations of either pulmonary fibrosis (e.g., as detected by chest radiography or pulmonary function tests) or cancer caused by asbestos exposures. Modern medical pulmonary imaging techniques or bioassays of circulating levels of cytokines or other biochemical factors associated with disease processes might be adaptable to better define early stages of asbestosis, and might provide a new paradigm for early detection of the active disease process. For example, positron emission tomographic imaging with use of tracers indicative of active collagen synthesis can detect fibrogenic response in a matter of weeks after quartz dust challenge in a rabbit animal model [Jones et al. 1997; Wallace et al. 2002].

3.5.3 Conduct Selective Epidemiological Studies of Workers Exposed to Asbestos Fibers and Other EMPs

Statistically powerful and well designed epidemiological studies are typically very expensive and time consuming, but they have been invaluable for defining associations between human health outcomes and occupational exposures. In fact, the strongest human evidence indicating that, at a sufficient dose and with a sufficient latency, certain EMPs of thoracic dimension and high durability pose risks for malignant and nonmalignant respiratory disease has come from epidemiological studies of workers exposed to asbestos fibers.

Outcomes from proposed research efforts outlined above in Section 3.5.2 may identify additional opportunities for informative epidemiological studies following the example of NIOSH researchers who have recently undertaken a reanalysis of data from a prior epidemiological study of asbestos textile workers, after having more thoroughly characterized exposures with use of sample filters archived from that study [Kuempel et al. 2006]. Outcomes from the approaches outlined above in Section 3.3.2 might also potentially identify opportunities for aggregate meta-analyses of data from multiple prior epidemiological studies, allowing an assessment of risks across various types of EMPs. Recently published research illustrates the potential of such meta-analyses to contribute to an understanding of determinants of disease caused by exposure to asbestos and related EMPs [Berman et al. 1995; Berman and Crump 2008b; Berman 2010].

Given the ongoing and widespread occupational and environmental exposure to Libby vermiculite, a more complete understanding of the mortality experience of the Libby occupational cohort could shed light on risks associated with exposure to the Libby amphibole, such as exposures at the World Trade Center disaster, as well as the health effects among the Libby community. Analyses of the Libby worker cohort continue and future analyses are envisioned, with the following aims:

- complete exposure-response modeling and occupational risk assessment for mesothelioma and asbestosis, and

- description of nonrespiratory outcomes (e.g., mortality with rheumatoid arthritis and mortality from extrapulmonary cancers).

Other research relating to Libby amphibole also continues. EPA and ATSDR have been engaged in a program of research involving several recent projects, including evaluation of

- the relationship between radiographic abnormalities and lung function in Libby community residents, finding that diffuse pleural thickening on radiography was a significant predictor of both restrictive and obstructive patterns on spirometry.

- the natural history of radiographic disease progression, observing an exposure-response relationship between cumulative fiber exposure and small opacity profusion level on chest radiographs of Libby workers.
- the effect of exposure to asbestos-containing Libby vermiculite at 28 processing sites in the United States. Activities included conducting medical screening of former workers and household contacts at six sites; a summary report is available at www.atsdr.cdc.gov/asbestos/sites/national_map.
- cases of mesothelioma, asbestosis, and lung cancer among former workers and others with nonoccupational exposure associated with a vermiculite processing facility in northeast Minneapolis.
- disease progression in workers exposed to asbestos-containing vermiculite ore at a fertilizer plant in Marysville, Ohio.
- autoimmune conditions not classically associated with asbestos exposure and health effects associated with low-level exposure and childhood exposure.

In addition, ATSDR continues to update its Tremolite Asbestos Registry (TAR) of individuals exposed to vermiculite-associated asbestiform amphibole in Libby. Opportunities for additional informative epidemiological studies relating to Libby amphibole could be pursued in the future, particularly if an EM-based job-exposure matrix for workers exposed to the Libby amphiboles is developed, or if amphibole exposures during commercial building and household construction renovation tasks were well-characterized.

Large unstudied populations with sufficiently high exposure to commercial asbestos fibers are unlikely to be identified in developed countries such as the United States, where asbestos use has been markedly curtailed and where occupational exposures have been strictly regulated in recent decades. Nevertheless, some developing countries (where asbestos use continues on a large scale and where exposures may be less regulated) may offer opportunities for *de novo* epidemiological studies that could contribute to a more refined understanding of the association of human health outcomes with occupational exposures to asbestos fibers and other EMPs.

Opportunities for epidemiological studies of exposed workers might be sought in other countries where medical registry data and historical or current workplace sampling data are available (e.g., in China, where epidemiological studies of another occupational dust disease, silicosis, have been collaboratively conducted by Chinese and NIOSH researchers [Chen et al. 2005]).

Opportunities may also exist in other countries for epidemiological studies of nonworker populations exposed to asbestos in ways not encountered in more developed countries. For example, regular whitewashing of the interiors of homes has, in more than one country, been shown to be fraught with hazard. In parts of Greece and Turkey and in New Caledonia, the local earthen material traditionally used for whitewashing homes was predominantly composed of tremolite asbestos, resulting in high rates of nonmalignant pleural plaques [Constantopoulos et al. 1987], lung cancer [Luce et al 2000; Menvielle et al. 2003], and malignant mesothelioma [Sakellariou et al. 1996; Senyiğit et al. 2000]. The whitewashing work, including crushing of the dry material before addition of water, was typically done by women with small children in tow, placing both sexes at risk of intermittent heavy exposures very early in life [Sakellariou et al. 1996]. This, along with the longer-term and lower-level exposures associated with inhabiting homes whitewashed with this asbestos-containing material, represents an exposure pattern very different from the occupational exposures to asbestos studied in the Unites States and other industrialized countries.

Results of epidemiological studies of workers exposed to EMPs, such as from nonasbestiform amphibole minerals or minerals in talc deposits, have provided limited evidence of an association between occupational exposure and lung cancer or mesothelioma. It will be important to establish *a priori* criteria to enable results of epidemiological studies or meta-analyses to be used to indicate whether or not occupational exposure to EMPs from nonasbestiform amphibole minerals or minerals in talc deposits is associated with a risk level that warrants preventive intervention. Clearly laying out these criteria and assessing the feasibility of conducting necessary studies should be done by a panel of knowledgeable experts. Laboratory research will undoubtedly shed much light on the issue of potential human health risks associated with specific physicochemical characteristics of EMPs. Still, where such studies are not only feasible but also likely informative, there is reason to consider further epidemiological studies of worker populations exposed to EMPs from nonasbestiform amphiboles. These may be conducted either *de novo* or through updating of prior studies for more complete follow-up of health outcomes and/or through reanalyzing archived exposure samples for development of more specific knowledge concerning etiologic determinants and quantitative risk.

Other studies worthy of consideration include these:

- Epidemiological studies of worker populations incidentally exposed to EMPs from fibrous minerals, including asbestiform minerals (e.g., those associated with Libby vermiculite);

- Epidemiological studies of populations exposed to other less-well-studied EMPs (e.g., wollastonite, attapulgite, and erionite); and

- Meta-analyses of data from multiple epidemiological studies of various worker populations, each exposed to EMPs with somewhat different attributes (such as EMP type or dimensions), to better define specific determinants of EMP-associated adverse health outcomes for purposes of risk assessment.

The following criteria should be considered in selecting and prioritizing possible populations for epidemiological study: (1) type of EMP exposure (e.g., mineral source, chemical composition, crystalline structure, surface characteristics, and durability); (2) adequate exposure information (e.g., EMP concentrations and [bivariate] EMP dimensions); (3) good work histories; (4) sufficient latency; (5) number of workers needed to provide adequate statistical power for the health outcome(s) of interest; and (6) availability of data on other potentially confounding risk factors. Priority should be placed on epidemiological studies with potential to contribute to the understanding of EMP characteristics that determine toxicity, including type of mineral source (e.g., asbestiform mineral habit vs. other fibrous mineral habit vs. blocky mineral habit) and morphology and other aspects of the airborne EMPs (e.g., dimensions [length and width], chemical composition, crystalline structure, surface characteristics, and durability).

In addition to epidemiological studies that address etiology and that quantify exposure-related risk, epidemiological studies can be used to better understand the pathogenesis of lung diseases caused by asbestos fibers and other EMPs. For example, appropriately designed or reconstructed epidemiological studies could be used to assess the relationship between the physicochemical properties of EMPs, lung fibrosis, and lung cancer.

3.5.4 Improve Clinical Tools and Practices for Screening, Diagnosis, Treatment, and Secondary Prevention of Diseases Caused by Asbestos Fibers and Other EMPs

Given the huge human and economic impact of asbestos-related disease and litigation, Congress has considered asbestos-related legislation on several occasions in recent years. To date, bills with provisions to require private industry to fund an asbestos victims' trust fund have not succeeded in Congress. Most recently, the Ban Asbestos in America Act, which was passed by the U.S. Senate in 2007 but was not acted on in the House of Representatives, would have authorized and funded a network of asbestos-related disease research and treatment centers to conduct studies, including clinical trials, on effective treatment, early detection, and prevention [U.S. Senate 2007]. This bill also called for the establishment of a mechanism for coordinating and providing data and specimens relating to asbestos-caused diseases from cancer registries and other centers, including a recently funded virtual biospecimen bank for mesothelioma [Mesothelioma Virtual Bank 2007].

Various research objectives relevant to clinical aspects of asbestos-related diseases are worthy of pursuit by NIOSH and other federal agencies, along with their partners, to improve screening, diagnosis, secondary prevention, and treatment. These include but are not limited to the following objectives:

- Continue to develop and validate technical standards for the assessment of digital chest radiographs with the ILO classification system. The ILO system for classifying chest radiographs of the pneumoconioses is widely used as a standard throughout the world. Although initially intended for use in epidemiological studies, the ILO system is now also commonly used as a basis for describing severity of disease in clinical care and for awarding compensation to individuals affected by nonmalignant diseases of the chest caused by asbestos and other airborne dusts. To ensure that digital chest radiographic methods used in future clinical and epidemiological studies can be compared with past studies based on conventional film radiography, there is a critical need to continue ongoing research to validate use of the ILO system for classification of digital chest images.

- Develop and promote standardized assessment of nonmalignant dust-induced diseases, including asbestos-related pleural and parenchymal disease, on computed tomography (CT) images of the chest. Over the past several decades, CT scanning of the chest has been increasingly used for assessing chest disease, and high-resolution CT scanning is often done in clinical settings. Although approaches for standardizing classifications of CT images for dust-related diseases have been proposed, none have yet been widely adopted or authoritatively promoted.

- Develop, validate, and promote standardization of approaches for assessment of past asbestos exposures by measurement of asbestos bodies and uncoated fibers, particularly in samples collected noninvasively (e.g., sputum). Various approaches for quantifying fiber burden have been used for research and clinical purposes, but results are often difficult or impossible to compare across different studies because of lack of standardization and the differential rates of biopersistence and translocation of various types of asbestos fibers.

- Develop and validate biomarkers for asbestosis, lung cancer, and mesothelioma to enable more specific identification of those at risk or early detection of disease in those previously exposed to asbestos. For example, noninvasive bioassays for mesothelioma warrant further research before they can be considered ready for routine application in clinical practice.

- Develop and/or adapt emerging medical imaging techniques to better define stages of asbestosis, or to provide a new paradigm for early detection or grading of the active disease process. For example, positron emission tomographic (PET) imaging using tracers indicative of active collagen synthesis can detect pulmonary fibrogenic response in a matter of weeks after quartz dust challenge in a rabbit animal model [Jones et al. 1997; Wallace et al. 2002]. This holds promise for noninvasive approaches for earlier clinical detection and more sensitive surveillance and epidemiological studies, that to date have been circumscribed by the long latency periods that characterize pulmonary fibrosis associated with asbestos exposure (e.g., as detected by conventional chest radiography).

- Develop new treatment options to reduce risk of malignant and nonmalignant disease among those exposed to asbestos and to effectively treat established asbestos-induced disease. For example, many widely used anti-inflammatory drugs exert their effect by inhibiting cyclooxygenase-2 (COX-2), an enzyme that is induced in inflammatory and malignant (including premalignant) processes. Promising results of laboratory and case-control epidemiological studies have led to clinical trials of COX-2 inhibitors as adjuvant therapy to enhance treatments for various types of cancer. Research is warranted to determine whether these drugs can reduce the risk of asbestos-related malignancies in exposed individuals.

- Authoritatively and regularly update and disseminate clear clinical guidance for practitioners, based on expert synthesis of the available literature.

3.6 Develop Improved Sampling and Analytical Methods for Asbestos Fibers and Other EMPs

There are important scientific gaps in understanding the health impacts of exposure to EMPs. Changes in how EMPs are defined for regulatory purposes will likely have to be accompanied by improvements to currently used analytical methods or development and application of new analytical methods. An ability to differentiate between fibers from the asbestos minerals and EMPs from their nonasbestiform analogs in air samples is an important need, especially for mineral-specific recommendations (e.g., occupational exposure limits). However, overcoming this obstacle may be difficult because of (1) the lack of standard criteria for the mineralogical identification of airborne EMPs and (2) technical difficulties in generating test aerosols of size-specific EMPs representative of worker exposures so that sampling and analytical methods can be tested and validated.

Improvements in exposure assessment methods are needed to increase the accuracy of the methods used to identify, differentiate, and count EMPs captured in air-sampling filter media. Until new analytical methods are developed and validated, it will be necessary to investigate the various proposals that have been made to modify current analytical methods, such as those discussed in Section 3.6.2, and additional modifications to the current analytical methods.

Manual microscopy methods are labor-intensive and error-prone. Automated analyses would permit examination of larger sample fractions and improve the accuracy of particle classification. Developing a practical method that accurately counts and sizes all EMPs could improve risk assessments and exposure assessments done in support of risk management. Automated methods could reduce operator bias and interlaboratory variability, providing more consistent results for risk assessments.

Some barriers to improving current analytical methods have been identified. Increasing the optical resolution of PCM analysis may help to increase counts of thinner asbestos fibers. However, any increases in optical microscopy resolution will not be sufficient to detect all asbestos fibers. In addition, any improvements in counting EMPs (e.g., an increase in the number of EMPs observed and counted) will need to be evaluated by comparing them with counts made by the current PCM method. The use of electron microscopy (EM) would improve the capability to detect thin fibers and also provide a means to identify many types of minerals.

However, the routine use of EM would

- require the development of standardized analytical criteria for the identification of various EMPs;
- require specialized experience in microscopy and mineral identification;
- increase analytical costs; and
- potentially increase the lag time between collecting the sample and obtaining results.

In some workplace situations, such as in construction, increases in the time needed to analyze samples and identify EMPs could potentially delay the implementation of appropriate control measures to reduce exposures.

Several potential sampling and analytical improvements are currently under study. Some of the studies are aimed at improving the accuracy of current techniques used for monitoring exposures to asbestos. One such study is evaluating the use of thoracic samplers for the collection of airborne EPs, and another concerns the use of gridded coverslips for PCM analyses. The proposed use of gridded coverslips for sample evaluation can aid in the counting of asbestos and other EMPs and can provide a means for "recounting" fibers at specific locations on a filter sample. Another study is evaluating the proposed ASTM method to determine whether interoperator variability of differential counting (to distinguish fibers of asbestos minerals from other EMPs) is within an acceptable range.

Research to develop new methods is warranted. One such research area could be development of methods that would permit assessment of the potential biopersistence (e.g., durability) of EMPs collected on air sampling filters prior to their evaluation by PCM or other microscopic methods. If durability is deemed biologically relevant, then an exposure assessment limited to the durable EMPs collected on a sample would help to reduce possible analytical interferences caused by other, nondurable EMPs and may eliminate the need for mineral identification. Another such area would be improvement in EM particle identification techniques, such as field emission SEM and the capability to determine the elemental composition of EMPs with an SEM equipped with EDS. A third research area, largely conceptual at this time, would be development of techniques to more accurately reflect internal dose by quantifying or estimating the number of particles likely to be produced by splitting of asbestos-fiber bundles deposited in the lung.

Modifications of current analytical methods and development of new analytical methods will require an assessment of their implications for worker health protection (e.g., how do the results using improved or new methods relate to human risk estimates based on counts of EMPs made by PCM?). To ensure that relevant toxicological parameters (e.g., dose, dimension, durability, and physicochemical parameters) are incorporated in the analysis and measurement, any changes in analytical methods should be made in concert with changes in how asbestos fibers or other EMPs are defined.

3.6.1 Reduce Interoperator and Interlaboratory Variability of the Current Analytical Methods Used for Asbestos Fibers

To ensure the validity of EMP counts on air samples, it is important to ensure consistency in EMP counts between and among analysts. Microscopy counts of EMPs on air sample filters are based on only a small percentage of the surface area of the filters, and the counting procedures require the analyst to make decisions on whether each observed particle meets specified criteria for counting. Interlaboratory sample exchange programs have been shown to be important for ensuring agreement in asbestos fiber counts between laboratories [Crawford et al. 1982]. Unfortunately, microscopists from different laboratories are unlikely to view exactly the same fields, and this alone accounts for some of the observed variation in fiber counts between microscopists. A mechanism to allow recounts of fibers from the exact same field areas would remove this variable and allow a better assessment of the variation attributable to microscopists in analyzing samples.

A technique is under development for improving the accuracy of PCM-based fiber-counting by allowing the same sample fields to be examined by multiple microscopists or by the same microscopist on different occasions [Pang et al. 1984, 1989; Pang 2000]. The method involves the deposition of an almost transparent TEM grid onto the sample. Included with the grid are coordinates that allow relocation of each grid opening. Photomicrographs of typical grid openings superimposed on chrysotile and amosite samples have been published [Pang et al. 1989]. Slides prepared in this manner have been used in a Canadian proficiency test program for many years. The main errors affecting the counts of various types of fibers (e.g., chrysotile, amosite, and SVF) have been evaluated by examining large numbers of slides by large numbers of participants in this program. A recently developed scoring system for evaluating the performance of microscopists is based on errors compared with a reference value defined for each slide by the laboratory in which they were produced [Pang 2002]. A statistical analysis of the intragroup precision in this study was able to identify those analysts who were outliers [Harper and Bartolucci 2003]. In a pilot study, the pooled relative standard deviations, without the outliers, met the requirements for an unbiased air-sampling method. Further study is needed to validate these findings and to identify other techniques that can reduce interlaboratory and interoperator variability in counting asbestos fibers and other EMPs by PCM.

Reference slides made from proficiency test filters from the American Industrial Hygiene Association (AIHA) have been created and circulated to laboratories and individual microscopists recruited from AIHA laboratory quality programs [Pang and Harper 2008; Harper et al. 2009]. The results illustrate an improved discrimination of fiber counts when the proficiency test materials have a more controlled composition. These reference slides have also been evaluated in Japan, the United Kingdom, and elsewhere in Europe. Further research will be useful in determining the value of these slides for training purposes.

3.6.2 Develop Analytical Methods with Improved Sensitivity to Visualize Thinner EMPs to Ensure a More Complete Evaluation of Airborne Exposures

Most PCMs can visualize EMPs with widths >0.25 µm, which is the approximate lower resolution limit when the microscope is operated at a magnification of 400× and calibrated to NIOSH 7400 specifications [NIOSH 1994a]. However, higher-end optical microscopes can resolve thinner widths, and, for crocidolite, they may resolve widths as thin as 0.1 µm.

Improvement in the optical resolution may be possible with use of an oil-immersion 100× objective with a numerical aperture of 1.49. Also, the use of 15× eyepiece oculars would help improve the visibility of small particles and thin EMPs in samples. However, using oil immersion has several drawbacks. When exposed to air for more than a few hours, the oil on the slide dries and its optical properties change. Also, the oil cannot be wiped off because the cover slip is likely to be moved and thus ruin the sample. For these reasons, using oil immersion does not permit recounts or further analysis for quality control purposes and is not an attractive alternative.

Other methods may also allow for increased resolution with optical microscopes. Anecdotal information on the use of PCM with dark-medium (DM) objectives, presented at a meeting in November 2007, suggests that analysts using DM objectives could resolve more blocks of the Health and Safety Executive/National Physical Laboratory (HSL/ULO) test slide‡‡ than are allowable for the method and produced higher counts of chrysotile fibers than expected [Harper et al. 2009]. The implication is that using DM objectives can resolve thinner chrysotile fibers than the accepted method. This methodology should be explored further to determine its resolution and potential application in asbestos exposure assessment.

As stated previously, because risk estimates for workers exposed to asbestos fibers have been

---

‡‡ The HSE/NPL Mark II or HSL/ULO Mark III Phase Shift Test Slide checks or standardizes the visual detection limits of the PCM. The HSL/ULO Test Slide consists of a conventional glass microscope slide with seven sets of parallel line pairs of decreasing widths. The microscope must be able to resolve the blocks of lines in accordance with the certificate accompanying the slide. Only slides where at least one block of lines is intended to be invisible should be used. Microscopes that resolve fewer or greater numbers of blocks than stated on the certificate cannot be used in NIOSH Method 7400.

based on counts made by the current PCM method, counts made with improved optical microscope resolution capabilities would not be directly comparable to current occupational exposure limits for airborne asbestos fibers. Additionally, the findings that asbestos fibers thinner than 0.1 μm are most associated with mesothelioma and that optical microscopes cannot resolve fibers <0.1 μm in width suggest that alternatives to PCM should be researched.

TEM can resolve asbestos fibers with widths less than ~0.01 μm, which effectively detects the presence of all asbestos fibers and other EMPs collected on airborne samples. Both TEM and SEM provide greater resolution for detecting and sizing EMPs. Both methods also provide capability for mineral identification (TEM using selected area X-ray diffraction [SAED], and TEM and SEM using EDS or WDS for elemental analysis). The cost of using TEM and/or SEM for routine analysis of all samples would be considerably higher than PCM analysis and the turnaround time for analysis would be substantially longer. In addition, any routine use of EM methods for counting and sizing asbestos fibers or other EMPs would require formal evaluation of interoperator and interlaboratory variability.

SEM is now a generally available method that can routinely resolve features down to ~0.05 μm, an order of magnitude better than optical microscopes. Field emission SEM is now commercially available and further increases this resolution. *In vitro* or short-term or long-term animal model studies can now utilize these EM imaging technologies to characterize EMPs for studies of etiology and disease mechanism. EM analyses of EMP size and composition can be supplemented with analysis of surface elemental composition by scanning Auger spectroscopy or X-ray photoelectron spectroscopy. Investigation is needed to determine whether SEM-backscatter electron diffraction analysis can be adapted to EMP crystallographic analyses equivalent to TEM-SAED capability. The ease of sample preparation and data collection for SEM analysis in comparison with TEM analysis, along with an SEM advantage in visualizing EMP and EMP morphology (e.g., surface characteristics), provides reason to reevaluate SEM methods for EMP characterization and mineral identification for field and laboratory sample analysis.

3.6.3 Develop a Practical Analytical Method for Air Samples to Differentiate Asbestiform Fibers from the Asbestos Minerals and EMPs from Their Nonasbestiform Analogs

A recently published ASTM method for distinguishing other EMPs from probable asbestos fibers uses PCM-determined morphologic features to differentiate asbestos fibers from other EMPs [ASTM 2006]. The proposed method has several points of deviation from existing PCM methods. It uses a new graticule that has not been tested for conformance with the traditional graticule used in standard PCM analysis of asbestos air samples. It specifies additional counting rules to classify particles, and there are few data to show these rules provide consistently achievable or meaningful results. Also, only limited data are available to show interoperator, intraoperator, or interlaboratory variation. These issues must be addressed before the method can be considered acceptable. NIOSH researchers are currently addressing these issues. Specific aims of the project are

- to determine the effect of using the traditional Walton-Beckett graticule and the new RIB graticule on the precision of measuring fiber dimensions; and

- to determine the interlaboratory variation of the proposed method for determining particle identities by observing morphological features of individual particles.

Anticipated outcomes of these ongoing research projects include a measure of method precision, which will help to determine whether the method meets the requirements of regulatory and other agencies.

Although EM may currently not be suitable for routine analysis of samples of airborne EMPs, EM techniques used to characterize and identify minerals (e.g., differentiating between asbestos fibers and other EMPs) should be further investigated and evaluated to determine whether results are reproducible by multiple microscopists and laboratories.

3.6.4 Develop Analytical Methods to Assess Durability of EMPs

Although research has been conducted to determine the ability of biological assays to evaluate the biopersistence of EMPs in the lung, there is a need to consider how the assessment of EMP durability might be incorporated into the evaluation of air samples containing a heterogeneous mix of EMPs. Research with several types of glass fibers and some other SVFs indicate that they dissolve in media at different rates depending on the pH and that they dissolve more rapidly than chrysotile and amphibole asbestos fibers [Leineweber 1984]. Chrysotile fibers have been shown to dissolve at a rate that varies not only with the strength of the acid, but also with the type of acid. Amphibole asbestos fibers have been shown to be more resistant to dissolution than chrysotile fibers. Research suggests that the rate of dissolution in the lungs for most EMPs appears to be strongly dependent on their chemical composition, surface characteristics, and dimension.

The selective dissolution of EMPs might be a useful approach for eliminating specific types of EMPs or other particles collected on air samples prior to analysis (e.g., microscopic counting). The removal of interfering EMPs prior to counting could potentially eliminate the need for additional analysis to identify EMPs on the sample. Selective dissolution of samples to remove interferences is well established in NIOSH practice for other analytes. NIOSH Method 5040 for diesel exhaust has an option for using acidification of the filter sample with hydrochloric acid to remove carbonate interference [NIOSH 2003a]. Silicate interferences for quartz by infrared spectroscopic detection are removed by phosphoric acid digestion in NIOSH Method 7603 [NIOSH 2003b]. Although selective dissolution might be accomplished for some EMPs, research will be necessary to develop and characterize a procedure that would correlate residual EMP counts to the results of toxicity studies.

3.6.5 Develop and Validate Size-selective Sampling Methods for EMPs

For measuring airborne concentrations of non-elongate particles in the workplace, conventions have been developed for sampling the aerosol fractions that penetrate to certain regions of the respiratory tract upon inhalation: the inhalable fraction of particulate that enters into the nose or the mouth; the thoracic fraction that penetrates into the thorax (i.e., beyond the larynx); and the respirable fraction that reaches the alveoli of the lung. The thoracic convention is recognized by NIOSH and other organizations that recommend exposure limits, and NIOSH has established precedence in applying it in RELs (e.g., the REL for metalworking fluid aerosols [NIOSH 1998]).

Asbestos fibers currently are collected for measurement by standard sampling and analytical methods (e.g., NIOSH Method 7400 [NIOSH 1994a], in OSHA ID–160 [OSHA 1998], in Methods for the Determination of Hazardous Substances (MDHS) 39/4 [HSE 1995], and in ISO 8672 [ISO 1993]). In these methods, air samples are taken by means of a membrane filter housed in a cassette with a cowled sampling head. Early studies [Walton 1954] showed that the vertical cowl excludes some very coarse particles because of elutriation, but its selection characteristics should have little effect on the collection efficiency for asbestos fibers. However, when Chen and Baron [1996] evaluated the sampling cassette with a conductive cowl used in sampling for asbestos fibers, they found inlet deposition was higher in field measurements than predicted by models.

Unlike the WHO [1997], NIOSH has not recommended an upper limit for width of asbestos fibers to be counted because airborne asbestos fibers typically have widths <3 μm. The absence of an upper width criterion for the NIOSH Method 7400 A rules has generated criticism that some EMPs counted by this method may not be of thoracic size. Others have recommended NIOSH Method 7400 B rules for the sampling and analysis of various types of fibers and EPs, including asbestos fibers [Baron 1996], because the B rules specify an upper limit of 3 μm for EP width. However, Method 7400 B rules have not been field-tested for occupational exposures to asbestos and many types of EPs.

Two separate but complementary investigations have examined the performance of thoracic samplers for EMPs [Jones et al. 2005; Maynard 2002].

Thoracic samplers allow the collection of airborne particles that meet the aerodynamic definition of thoracic-size EMPs (i.e., with physical widths equal to or less than 3 μm for the typical length distributions of fibers of silicate composition), collecting only those EMPs considered most pathogenic. The results of studies have indicated that penetration of some thoracic samplers is independent of EMP length, at least up to 60 μm, indicating that the samplers' penetration characteristics for an EP aerosol should be no different than that of an isometric aerosol. In a study be Jones et al. [2005], the relative ability of the thoracic samplers to produce adequately uniform distributions of EPs on the surface of the membrane filter was also tested. On the basis of results of these studies, two samplers appeared to meet the criteria of minimal selection bias with respect to EP length and uniform distribution on the collection filters. However, neither of these samplers has been tested under conditions of field use. NIOSH is currently evaluating these two thoracic samplers and the traditional cowled sampler in three different mining environments. The results from these studies have been published [Lee et al. 2008; Lee et al. 2010], and each of the samplers gave fiber concentrations in proportion to their fiber loading rates, which is consistent with previous reports [Cherrie et al. 1986]. Thus, alternative samplers cannot be recommended until the impact of this effect on the established risk analysis has been evaluated.

3.7 From Research to Improved Public Health Policies for Asbestos Fibers and Other EMPs

Section 3 of this *Roadmap* proposes several strategic goals and associated objectives for a multidisciplinary research program on asbestos fibers and other EMPs. In summary, accomplishing these goals is intended to (1) further elucidate the physicochemical properties that contribute to their pathogenicity; (2) improve existing analytical tools and develop new analytical tools for identifying and measuring exposures to EMPs with use of metrics that reflect the important determinants of toxicity (e.g., dimension and composition); (3) better define the nature and extent of occupational exposures to EMPs and their relationships to EMP-related health outcomes among exposed worker populations; and (4) improve clinical tools for screening, diagnosis, secondary prevention, and treatment of EMP-related diseases.

Results of much of the research to date (e.g., animal and human studies with asbestos and other EMPs) are readily available and should be considered in developing the research program, including the specification of minerals to be studied. Much of this evidence supports the important role of particle dimension as a determinant of lung deposition and retention and the concomitant role of particle composition and crystalline structure as a determinant of durability and biopersistence. Despite this body of research, several fundamental issues are not clearly understood, and a broad systematic approach to further toxicological and epidemiological research would help to reduce remaining uncertainties. Although long, thin asbestos fibers clearly cause respiratory disease, the role of unregulated short (i.e., <5 μm) asbestos fibers is not entirely clear. It also remains unclear to what extent each of the various physicochemical parameters of asbestos fibers is responsible for respiratory disease outcomes (e.g., asbestosis, lung cancer, and mesothelioma) observed in asbestos-exposed individuals. Limited evidence from studies with other EMPs confirms the importance of particle dimension and biopersistence in causing a biological response. However, uncertainty remains as to whether the respiratory disease outcomes for workers exposed to asbestos fibers can be anticipated for workers exposed to other EMPs of thoracic size and with elemental compositions similar to asbestos.

Another important effort that can inform development of the research program will involve a systematic collection and review of available information on (1) industries and occupations with exposure to EMPs; (2) airborne exposure in these industries and occupations; and (3) numbers of workers potentially exposed in these industries and occupations. Additional relevant minerals and mineral habits identified should also be considered for study. The minerals identified through these efforts should be carefully and comprehensively characterized with respect to both structure and elemental composition. In the characterization of minerals, consideration should also be given to (1) purity, (2) particle morphology (range of dimensions and sizes), (3) surface area, (4) surface chemistry, and (5) surface reactivity. Care must be taken to ensure that a sufficient amount of the studied material is available, not only for current studies but also as reference material for possible future studies. The information developed from all of these efforts should be entered into a database, which can serve as a tool for selection of minerals for testing and validation of toxicological tests, as well as to assist in identification of worker populations for possible epidemiological studies.

An objective of the proposed research is to achieve a level of mechanistic understanding that can provide a basis for developing biologically based models for extrapolating results of animal inhalation and other types of *in vivo* studies to predict risks to worker health associated with exposure conditions typically encountered in workplaces. Little is known about the mechanisms by which asbestos fibers and some other EMPs cause lung cancer, mesothelioma, and nonmalignant respiratory diseases. As these mechanisms become understood, biologically based models can be developed to extrapolate from exposure-dose-response relationships observed in animals to estimates of disease risk in exposed humans. In addition, such studies would provide (1) an opportunity to measure molecular and cellular outcomes that can be used to determine why one animal species responds differently from another and (2) information on EMP characteristics associated with eliciting or potentiating various biological effects. The outcomes of these studies can then be evaluated in subsequent experiments to provide (1) risk assessors with a useful understanding of the various disease mechanisms by which animals respond to EMP exposures; and (2) regulatory agencies and industrial hygiene and occupational health professionals with information needed to implement appropriate exposure limits and risk management programs for monitoring worker exposure and health.

It is anticipated that it may be difficult to find populations of workers that are exposed to EMPs with characteristics (e.g., dimension, composition) of interest that are sufficiently large to provide adequate statistical power, and where exposures are not confounded or where confounding can be effectively controlled in the analysis. NIOSH retains exposure information and, in some cases, personal air sample filters collected and archived from past epidemiological studies of workers exposed to asbestos. Such existing data might be used to update and extend findings from these studies. Where appropriately balanced epidemiological studies can be identified, it may be possible to conduct meta-analyses to investigate important EMP characteristics. The analysis of archived samples may help to elucidate how more detailed characteristics of exposure (e.g., particle dimension) relate to disease outcomes. New epidemiological (retrospective and prospective) studies should not be undertaken unless feasibility studies (e.g., preliminary assessments of study population size, exposure latencies, records of exposure, and confounders) have been appropriately considered.

Because the opportunities for informative epidemiological studies are likely to be limited, it will be necessary to complement them with toxicological testing, and an integrated approach to toxicological research will be needed to understand how various types of EMPs induce disease. Where epidemiological studies of new cohorts are possible, or where epidemiological studies of previously studied cohorts can be updated, attempts should be made to link their results with those of toxicological studies to assess the ability of various types of toxicological testing to predict health outcomes in humans. Toxicological testing should be done with attention to collecting more specific information, including: (1) physical characteristics (e.g., dimension); (2) chemical composition; (3) *in vitro* acellular data (dissolution, durability); and (4) *in vitro/in vivo* cellular data (e.g., cytotoxicity, phagocytosis, chromosomal damage, mediator release).

To help elucidate which physicochemical properties are important for inducing a biological effect, it may be necessary to generate exposures to EMPs of specific dimensions and composition. Several approaches are being pursued by NIOSH to overcome technological difficulties in generating sufficient quantities of well-characterized and dimensionally restricted EMPs. Efforts to generate mineral samples of appropriate particle-size dimensions by grinding techniques have met with some success but have not consistently generated EMPs in restricted size ranges of interest or in sufficient quantity to enable toxicity testing. Another approach has used a fiber size classifier [Deye et al. 1999], but this has not provided large enough quantities of EMPs for long-term inhalational exposure studies in animals. NIOSH researchers are currently evaluating the possibility of developing a fiber size classifier with increased output to generate much larger quantities of particles in restricted size ranges for toxicological testing.

An outcome of the proposed research programs should be an understanding of the relationships between and among the results of human observational studies and *in vitro*, short-term *in vivo*, and long-term *in vivo* experimental studies. Any research undertaken should be designed to ensure that results can be interpreted and applied within the context of other studies. For example, EMPs used in long-term animal inhalation studies should also be tested in *in vitro/in vivo* assay systems so that findings can be compared. The results of such experiments can help to develop and standardize *in vitro/in vivo* assay systems for use in predicting the potential toxicity of various types of EMPs.

Government agencies, other organizations, and individual researchers have already recommended similar research strategies for evaluating the toxicity of mineral and synthetic fibers [Greim 2004; ILSI 2005; Mossman et al. 2007; Schins 2002; Vu et al. 1996]. These published strategies should be used as a foundation for developing a research program.

Some research and improvements in sampling and analytical methods used to routinely assess exposures to EMPs can be done in the short term, and because the results of the toxicological studies provide a clearer understanding of EMP characteristics that determine toxicity, it will be necessary to ensure that the measurement techniques used in evaluating workplace exposures incorporate the exposure metrics used in determining the dose-response effect found in animal studies. The development of such exposure measurement techniques should (1) reduce the subjectivity inherent in current methods of particle identification and counting, (2) closely quantify EMPs on the basis of characteristics that are important to toxicity, and (3) reduce cost and shorten turnaround times in comparison with current EM methods.

Toxicological, exposure assessment, and epidemiological research should be conducted with the overarching goal of developing information necessary for risk assessments. Although a framework for mineralogical research is not provided in this *Roadmap*, research providing a better understanding of mineralogical physicochemical properties will help inform the risk assessment process. Improved risk assessments and analytical methodology are needed to inform the development of new and revised occupational exposure limits for control of exposures associated with the production of EMP-caused disease.

For those individuals who have an asbestos-related disease or are at high risk of developing an asbestos-related disease, research is needed to improve methods and clinical guidance for screening, diagnosis, secondary prevention, and treatment of EMP-caused diseases. The development and validation of biomarkers of disease and improved lung-imaging technologies can lead to earlier diagnosis of asbestos-related disease. It will also be important to advance knowledge on how to effectively treat EMP-caused diseases, especially malignant mesothelioma, which is fatal in most cases. Accomplishing the goals of early diagnosis and development of treatment options can improve the quality and quantity of life for those who develop asbestos-related disease.

4 The Path Forward

Developing an interdisciplinary research program and prioritizing research projects to implement the research agenda envisioned in this *Roadmap* will require a substantial investment of time, scientific talent, and resources by NIOSH and its partners. However, achieving the proposed goals will be well worth the investment because it will improve the quality of life of U.S. workers by preventing workplace exposure to potentially hazardous EMPs, and it will reduce future healthcare costs. As with any strategic approach, unintended and unforeseen results and consequences will require program adjustments as available resources change, information is produced, and time goes on.

4.1 Organization of the Research Program

To ensure that the scientific knowledge created from implementation of this *Roadmap* is applied as broadly as possible, NIOSH plans to partner with other federal agencies, such as the Agency for Toxic Substances and Disease Registry (ATSDR), the Consumer Product Safety Commission (CPSC), the Environmental Protection Agency (EPA), the Mine Safety and Health Administration (MSHA), the National Institute of Standards and Technology (NIST), the National Institute of Environmental Health Sciences (NIEHS), the National Toxicology Program (NTP), the Occupational Safety and Health Administration (OSHA), and the United States Geological Survey (USGS), as well as with labor, industry, academia, practitioners, and other interested parties including international groups. Partnerships and collaborations will be used to help focus the scope of the research to be undertaken, enhance extramural research activities, and assist in the development and dissemination of educational materials describing the outcomes of the research and their implications for occupational and public health policies and practices.

Some of the next steps in development will involve organizing study groups with representatives from federal agencies, industry, academia, and workers' organizations to identify specific priorities for the programs developed within the overarching research framework. Study groups should be assembled from among the partners to identify specific research elements needed to address the information gaps and data needs outlined in this *Roadmap*. Although it may be appropriate to organize separate study groups around the scientific disciplines needed to conduct the research—such as epidemiology, toxicology, exposure assessment, mineralogy, and particle characterization and analysis—and to conduct the risk assessments, each of the study groups will need to include members from other disciplines to ensure the multidisciplinary nature of the research is considered and addressed. Also important will be coordination between and among study groups to ensure the efforts in the various research areas are complementary and move toward common goals and the eventual development of sufficient information for risk assessment. These study groups should be maintained over the lifetime of the research program to oversee and help guide the research. An independent group could provide oversight of the overall research effort, periodically reviewing the various discipline-specific research programs to help ensure that the most appropriate research is accomplished in a timely and coordinated manner and to help maintain the scientific quality of the research.

4.2 Research Priorities

The key issues discussed in Section 2.10 include several research needs: (1) for the asbestos minerals, development of a clearer understanding of the important dimensional and physicochemical determinants of pathogenicity; (2) for other EMPs, such as those from nonasbestiform habits of the asbestos minerals and erionite, development of a deeper understanding of the determinants of toxicity; and (3) development of analytical methods that can differentiate EMPs and quantify airborne exposures to EMPs. To begin addressing these issues, infrastructure projects should be developed and initiated with input from the study groups.

One such infrastructure project is the development of a standardized set of terms that can be used to clearly and precisely describe minerals and other scientific concepts. This is needed to help with the planning of research projects and to effectively communicate research results. This effort should involve representatives from each of the relevant scientific disciplines.

Another infrastructure project that should be considered at the onset of prioritizing research is the development of criteria and logistics for establishing a mineral reference repository. Initially, representative samples from the known asbestos deposits should be procured and carefully and comprehensively characterized. If samples of these repository minerals are further processed in the course of conducting research, the processed materials will need to be fully characterized as well. Concomitant with this characterization effort should be the development of a mineralogical research effort addressing issues pertaining to the identification of minerals that might be found on airborne samples collected at various workplace environments and to develop further and deeper understanding of mineralogical properties that may contribute to the toxicity of particles.

One of the earliest research efforts will be preliminary systematic collection and evaluation of available information on (1) industries/occupations/job tasks/processes with exposure to various types of asbestos fibers and other EMPs, (2) numbers of workers exposed, (3) characteristics and levels of exposures, and (4) associated particulate exposures. The knowledge generated from these efforts will be needed to identify the EMPs that workers are exposed to and worker populations that have the potential to be included in epidemiological studies. In addition to ascertaining EMP exposures and EMP-exposed populations in the United States, networking and other tools should be used to identify potential international populations for epidemiological studies. Representative samples of the EMPs identified through these efforts should be procured, characterized, and included in the mineralogical reference repository. After thorough characterization, these samples can be classified and prioritized for use in the toxicological studies.

A part of this early effort should be the development of a comprehensive and integrated public-use information management system to warehouse (1) the mineral characterization information generated on the reference samples; (2) data generated from hazard and health surveillance activities; (3) information on the minerals tested and the methods used, as well as the results of toxicological studies; and (4) the data gathered from epidemiological and other surveillance investigations. By having the results of previous studies available in the information management system, it could be used to promote the development of an efficient, non-duplicative research program. It could also be a resource for data exploration and additional analyses of accumulated results.

After comprehensive review of current knowledge and the available data in the above-described information management system, the study groups should identify specific aims and plan, prioritize, and conduct mineralogical, toxicological, epidemiological, and clinical research within the general framework laid out in this *Roadmap*. Early results will inform the need for later research and will dictate changes in priorities and directions needed to accomplish the overall goals of the research program.

Ongoing research and study on improvements of the analytical methods currently used for regulatory purposes should be independent of other research. However, as surveillance and exposure assessment efforts proceed, research on analytical methods should advance the capability to identify and characterize worker exposures and to measure relevant exposure parameters identified by toxicological research. Eventually, after determinants of EMP toxicity are more fully elucidated, research should increasingly focus on sampling and analytical methods that can be routinely used in compliance exposure assessment.

4.3 Outcomes

NIOSH will promote integration of the research goals set forth in this *Roadmap* into the industry sector–based and research-to-practice-focused National Occupational Research Agenda (NORA), a national program involving both the public and private sectors. The goals and objectives of this *Roadmap* can be substantially advanced through robust public-private sector partnership.

The ideal outcome of a comprehensive research program for asbestos fibers and other EMPs would be to use the results to develop recommendations to protect workers' health that are based on unambiguous science. Optimally, such recommendations may specify criteria, such as a range of chemical composition, dimensional attributes (e.g., ranges of length, width, and aspect ratio), dissolution rate/fragility parameters, and other factors that can be used to indirectly assess the toxicity of EMPs. It would be particularly advantageous if the results of the research could be used to devise a battery of validated *in vitro* or short-term *in vivo* assays with sufficient predictive value to identify EMPs warranting concern on the basis of their physical and chemical properties, without the need for comprehensive toxicity testing and/or epidemiological evaluation of each individual EMP. Newly identified EMPs could be compared to the criteria to determine a likelihood of toxicity. Coherent risk management approaches for EMPs that fully incorporate a clear understanding of the toxicity could then be developed to minimize the potential for EMP-related disease outcomes among exposed workers.

Although it is beyond the scope of this *Roadmap*, exploration of the extent to which a health-protective policy for EMPs could be extended to SVFs and other manufactured materials, such as engineered carbon nanotubes and nanofibers, is warranted. It has been noted that elongate nanoscale particles (e.g., single-walled and multiwalled carbon nanotubes) cause interstitial fibrosis in mice [Shvedova et al. 2005; Porter et al. 2009] and that peritoneal exposure of mice to carbon nanotubes has been reported to induce pathological responses similar to those caused by asbestos, suggesting potential for induction of mesothelioma [Poland et al. 2008]. Recommendations have been made elsewhere to systematically investigate the health effects of these manufactured nanomaterials within the next five years [Maynard et al. 2006; NIOSH 2008b]. Integrating results of nanoparticle toxicity investigations with the results of the research program developed as a result of this *Roadmap* may lead to a broader and more fundamental understanding of the determinants of toxicity of EPs.

Working toward achieving the goals delineated in this *Roadmap* is consonant with NIOSH's statutory mission to generate new knowledge in the field of occupational safety and health and to transfer that knowledge into practice for the benefit of workers. Advancing knowledge relevant for use in protecting workers from adverse health effects arising from exposure to asbestos fibers and other EMPs is the ultimate goal. Though further scientific research conducted by NIOSH will continue to focus on the occupational environment, NIOSH intends to pursue partnerships to ensure that scientific research arising from this *Roadmap* will comprise an integrated approach to understanding and limiting EMP hazards incurred not only in work settings, but also in the general community and the general environment.

As information becomes available from research activities, NIOSH plans to assist in the development and dissemination of educational materials describing the outcomes of the research and their implications for occupational and public health policies and practices.

5 References

Aadchi S, Yoshida S, Kawamura K, Takahashi M, Uchida H, Odagirl Y, Taekmoto K [1994]. Induction of oxidative DNA damage and mesothelioma by crocidolite with special reference to the presence of iron inside and outside of asbestos fiber. Carcinogenesis 15:753–758.

Abrams D, Fernandez F, Bosl J, Ruiz W [2010]. Letter communication to the NIOSH Docket Office for Docket No. NIOSH-099C [http://www.cdc.gov/niosh/docket/archive/pdfs/NIOSH-099C/0099C-041410-Abrams_et_al_sub.pdf]. Date accessed: August 19, 2010.

Addison J [2007]. Letter communication to the NIOSH Docket Office for Docket No. NIOSH-099 [http://www.cdc.gov/niosh/docket/pdfs/NIOSH-099/0099-052707-addison_sub.pdf]. Date accessed: June 30, 2008.

Addison J, McConnell EE [2008]. A review of carcinogenicity studies of asbestos and non-asbestos tremolite and other amphiboles. Regul Toxicol Pharmacol 52(Suppl 1):S187–S199.

Aderem A [2002]. How to eat something bigger than your head. Cell 110:5–8.

Allison AC, Ferluga J [1977]. Cell membranes in cytotoxicity. Adv Exp Med Biol 84:231–246.

Amandus HE, Wheeler R [1987]. The morbidity and mortality of vermiculite miners and millers exposed to tremolite–actinolite. Part II: mortality. Am J Ind Med 11:15–26.

Amandus HE, Althouse R, Morgan WK, Sargent EN, Jones R [1987a]. The morbidity and mortality of vermiculite miners and millers exposed to tremolite-actinolite. Part III: radiographic findings. Am J Ind Med 11:27–37.

Amandus HE, Wheeler R, Jankovic J, Tucker J [1987b]. The morbidity and mortality of vermiculite miners and millers exposed to tremolite-actinolite. Part I: exposure estimates. Am J Ind Med 11:1–14.

Neuendorf KKE, Mehl JP, Jackson JA, eds [2005]. Glossary of geology. 5th ed. Alexandria, VA: American Geological Institute.

Anonymous [1997]. Asbestos, asbestosis, and cancer: the Helsinki criteria for diagnosis and attribution. Scand J Work Environ Health 23:311–316.

Ansari FA, Ahmad I, Ashqauin M, Yunus M, Rahman Q [2007]. Monitoring and identification of airborne asbestos in unorganized sectors, India. Chemosphere 68:716–723.

Asgharian B, Yu CP [1988]. Deposition of inhaled fibrous particles in the human lung. J Aerosol Med 1:37–50.

ATS (American Thoracic Society) [2004]. Diagnosis and initial management of nonmalignant diseases related to asbestos. Am J Respir Crit Care Med 170:691–715 [http://www.thoracic.org/sections/publications/statements/pages/eoh/asbestos.html]. Date accessed: June 30, 2008.

ASTM (American Society of Testing Materials) [2006]. Work item WK3160: new standard test method for sampling and counting airborne fibers, including asbestos fibers, in mines and quarries, by phase contrast microscopy. West Conshohocken, PA: ASTM International.

ATSDR (Agency for Toxic Substances and Disease Registry) [2001]. Toxicological profile for asbestosis [http://www.atsdr.cdc.gov/toxprofiles/tp61.html]. Date accessed: January 26, 2007.

ATSDR (Agency for Toxic Substances and Disease Registry) [2003]. Report on the expert panel on health effects of asbestos and synthetic vitreous fibers: the influence of fiber length. Prepared by Eastern Research Group [http://www.atsdr.cdc.gov/HAC/asbestospanel/]. Date accessed: January 26, 2007.

Aung W, Hasegawa S, Furukawa T, Saga T [2007]. Potential role of ferritin heavy chain in oxidative stress and apoptosis in human mesothelial and mesothelioma cells: implications for asbestos-induced oncogenesis. Carcinogenesis 28:2047–2052.

Axelson O, Steenland K [1988]. Indirect methods of assessing the effects of tobacco use in occupational studies. Am J Ind Med 13:105–118.

Axelson O [1989]. Confounding from smoking in occupational epidemiology. Br J Ind Med 46:505–507.

Bain GW [1942]. Vermont talc and asbestos deposits. In: Newhouse WH, ed. Ore deposits as related to structural features. Princeton, NJ: Princeton University Press, pp. 255–258.

Bang KM, Pinheiro GA, Wood JM, Syamlal G [2006]. Malignant mesothelioma mortality in the United States, 1999–2001. Int J Occup Environ Health 12:9–15.

Baris YI, Grandjean P [2006]. Prospective study of mesothelioma mortality in Turkish villages with exposure to fibrous zeolite. J Natl Cancer Inst 98:414–417.

Bariş B, Demir AU, Shehu V, Karakoca Y, Kisacik G, Bariş YI [1996]. Environmental fibrous zeolite (erionite) exposure and malignant tumors other than mesothelioma. J Environ Pathol Toxicol Oncol 15:183–189.

Baris YI, Saracci R, Simonato L, Skidmore JW, Artvinli M [1981]. Malignant mesothelioma and radiological chest abnormalities in two villages in Central Turkey: an epidemiological and environmental investigation. Lancet 1:984–987.

Baron P [1996]. Application of the thoracic sampling definition to fiber measurement. Am Ind Hyg Assoc J 57:820–824.

Barrett CJ [1994]. Cellular and molecular mechanisms of asbestos carcinogenicity: implications for biopersistence. Environ Health Perspect 102:19–23.

Beckett ST, Jarvis JL [1979]. A study of the size distribution of airborne amosite fibers in the manufacture of asbestos insulating boards. Ann Occup Hyg 22:273–284.

Beard ME, Crankshaw OS, Ennis JT, Moore CE [2001]. Analysis of crayons for asbestos and other fibrous materials, and recommendations for improved analytical definitions. Research Triangle Park, NC: Research Triangle Institute, Center for Environmental Measurements and Quality Assurance, Earth and Mineral Sciences Department, pp. 23.

Bellmann B, Muhle H, Pott F, Konig H, Kloppel H, Spurny K [1987]. Persistence of man-made mineral fibres (MMMF) and asbestos in rat lungs. Ann Occup Hyg 31:693–709.

Bergstrand H [1990]. The generation of reactive oxygen-derived species by phagocytes. Agents Actions Suppl 30:199–211.

Berman DW, Crump KS [2008a]. Update of potency factors for asbestos-related lung cancer and mesothelioma. Crit Rev Toxicol 38(Suppl 1):1–47.

Berman DW, Crump KS [2008b]. A meta-analysis of asbestos-related cancer risk that addresses fiber size and mineral type. Crit Rev Toxicol 38(Suppl 1):49–73.

Berman DW [2010]. Comparing milled fiber, Quebec ore, and textile factory dust: has another piece of the asbestos puzzle fallen into place? Crit Rev Toxicol 40:151–188.

Berman DW, Crump KS, Chatfield EJ, Davis JMG, Jones AD [1995]. The sizes, shapes, and mineralogy of asbestos structures that induce lung tumors or mesothelioma in AF/HAN rats following inhalation. Risk Anal 15:181–195.

Bernstein DM, Morscheidt C, Grimm HG, Thevenaz P, Teichert U [1996]. Evaluation of soluble fibers using the inhalation biopersistence model, a nine-fiber comparison. Inhal Toxicol 8:345–385.

Bernstein DM, Sintes JMR, Ersboell BK, Kunert J [2001]. Biopersistence of synthetic mineral fibers as a predictor of chronic inhalation toxicity in rats. Inhal Toxicol 13:823–849.

Bernstein DM, Donaldson K, Decker U, Gaering S, Kunzendorf P, Chevalier J, Holm SE [2008]. Biopersistence study following exposure to chrysotile asbestos alone or in combination with fine particles. Inhal Toxicol 20:1009–1028.

Bertino P, Marconi A, Palumbo L, Bruni BM, Barbone D, Germano S, Dogan AU, Tassi GF, Porta C, Mutti L, Gaudino G [2007]. Erionite and asbestos differently cause transformation of human mesothelial cells. Int J Cancer 121:12–20.

Blair A, Stewart P, Lubin JH, Forastiere F [2007]. Methodological issues regarding confounding and exposure misclassification in epidemiological studies of occupational exposures. Am J Ind Med 50:199–207.

Blake T, Castranova V, Schwegler-Berry D, Baron P, Deye GJ, Li C, Jones W [1998]. Effect of fiber length on glass microfiber cytotoxicity. J Toxicol Environ Health 54:243–259.

Boettcher AL [1966]. The Rainy Creek igneous complex near Libby, Montana [PhD dissertation]. University Park, PA: The Pennsylvania State University.

Bonneau L, Malard C, Pezerat H [1986]. Role of dimensional characteristics and surface properties of mineral fibers in the induction of pleural tumors. Environ Res 41:268–275.

Brain JD, Godleski J, Kreyling W [1994]. *In vivo* evaluation of chemical biopersistence of nonfibrous inorganic particles. Environ Health Perspect 102(Suppl 5):119–125.

British Thoracic Society Standards of Care Committee [2001]. Statement on malignant mesothelioma in the United Kingdom. Thorax 56:250–265.

Brody AR, Hill LH [1983]. Interactions of chrysotile asbestos with erythrocyte membranes. Environ Health Perspect 51:85–89.

Brown BM, Gunter ME [2003]. Morphological and optical characterization of amphiboles from Libby, Montana, U.S.A. by spindle-stage assisted polarized light microscopy. Microscope 51:121–140.

Brown DM, Beswick PH, Donaldson K [1999]. Induction of nuclear translocation of NF-κB in epithelial cells by respirable mineral fibres. J Pathol 189:258–264.

Brown DP, Dement JM, Wagoner JK [1979]. Mortality patterns among miners and millers occupationally exposed to asbestiform talc. In: Lemen R, Dement J, eds. Dusts and disease: occupational and environmental exposures to selected fibrous and particulate dusts. Park Forest South, IL: Pathotox Publishers, pp. 317–324.

Brown DP, Sanderson W, Fine LJ [1990]. NIOSH health hazard evaluation report: R.T. Vanderbilt Company, Gouverneur, New York. HETA 90-390-2065, MHETA 86-012-2065 [http://www.cdc.gov/niosh/hhe/reports/pdfs/1990-0390-2065.pdf]. Date accessed: June 30, 2008.

Brown DP, Kaplan SD, Zumwalde RD, Kaplowitz M, Archer VE [1986]. Retrospective cohort mortality study of underground gold mine workers. In: Goldsmith D, Winn D, Shy C, eds. Silica, silicosis, and lung cancer. New York: Praeger, pp. 335–350.

Brunner WM, Williams AN, Bender AP [2008]. Investigation of exposures to commercial asbestos in northeastern Minnesota iron miners who developed mesothelioma. Regul Toxicol Pharmacol 52(Suppl 1):S116–S120.

Cady WM, Albee AL, Chidester AH [1963]. Bedrock geology and asbestos deposits of the upper Missisquoi Valley and vicinity, Vermont. U.S. Geological Survey Bulletin 1122–B, 78 pp, 1 plate.

Campbell WJ, Steel EB, Virta RL, Eisner MH [1979]. Relationship of mineral habit to size characteristics of tremolite cleavage fragments and fibers. Report of Investigations #8367. Washington, DC: U.S. Department of the Interior, Bureau of Mines.

Carbone M, Bedrossian CW [2006]. The pathogenesis of mesothelioma. Semin Diagn Pathol 23:56–60.

Carbone M, Kratzke RA, Testa JR [2002]. The pathogenesis of mesothelioma. Semin Oncol 29:2–17.

Cardinali G, Kovacs D, Maresca V, Flori E, Dell'Anna ML, Campopiano A, Casciardi S, Spagnoli G, Torrisi MR, Picardo M [2006]. Differential *in vitro* cellular response induced by exposure to synthetic vitreous fibers (SVFs) and asbestos crocidolite fibers. Exp Mol Pathol 81:31–41.

CFR. Code of Federal regulations. Washington, DC: U.S. Government Printing Office of the Federal Register.

Chen C, Baron PA [1996]. Aspiration efficiency and wall deposition in the fiber sampling cassette. Am Ind Hyg Assoc J 57:142–152.

Chen W, Hnizdo E, Chen JQ, Attfield MD, Gao P, Hearl F, Lu J, Wallace WE [2005]. Risk of silicosis in cohorts of Chinese tin and tungsten miners, and pottery workers (I): an epidemiological study. Am J Ind Med 48:1–9.

Cherrie J, Jones AD, Johnston AM [1986]. The influence of fiber density on the assessment of fiber concentration using the membrane filter method. AIHAJ 47:465–474.

Chisholm JE [1973]. Planar defects in fibrous amphiboles. J Mater Sci 8:475–483.

Churg A, Stevens B [1995]. Enhanced retention of asbestos fibers in the airways of human smokers. Am J Respir Crit Care Med 151:1409–1413.

Churg A, Wright JL, Hobson J, Stevens B [1992]. Effects of cigarette smoke on the clearance of short asbestos fibres from the lung and a comparison with the clearance of long asbestos fibres. Int J Exp Pathol 73:287–297.

Coffin DL, Palekar LD, Cook PM [1982]. Tumorigenesis by a ferroactinolite mineral. Toxicol Lett 13:143–150.

Coffin DL, Cook PM, Creason JP [1992]. Relative mesothelioma induction in rats by mineral fibers: comparison with residual pulmonary mineral fiber number and epidemiology. Inhal Toxicol 4:273–300.

Comba P, Gianfagna A, Paoletti L [2003]. Pleural mesothelioma cases in Biancavilla are related to a new fluoro-edenite fibrous amphibole. Arch Environ Health 58:229–232.

Constantopoulos SH, Saratzis NA, Kontogiannis D, Karantanas A, Goudevenos JA, Katsiotis P [1987]. Tremolite whitewashing and pleural calcifications. Chest 92:709–712.

Cook PM, Palekar LD, Coffin DL [1982]. Interpretation of the carcinogenicity of amosite asbestos and ferroactinolite on the basis of retained fiber dose and characteristics in vivo. Toxicol Lett 13:151–158.

Cooper WC, Wong O, Graebner R [1988]. Mortality of workers in two Minnesota taconite mining and milling operations. J Occup Med 30:506–511.

Cooper WC, Wong O, Trent LS, Harris F [1992]. An updated study of taconite miners and millers exposed to silica and nonasbestiform amphiboles. J Occup Med 34:1173–1180.

Crawford NP, Thorpe HL, Alexander W [1982]. A comparison of the effects of different counting rules and aspect ratios on the level and reproducibility of asbestos fiber counts. Part I: effects on level. Edinburgh, UK: Institute of Occupational Medicine.

Cullen MR [2005]. Serum osteopontin levels: is it time to screen asbestos-exposed workers for pleural mesothelioma? N Engl J Med 353:1617–1618.

Cullen RT, Miller BG, Davis JMG, Brown DM, Donaldson K [1997]. Short-term inhalation and in vitro tests as predictors of fiber pathogenicity. Environ Health Perspect 105(Suppl):1235–1240.

Cummins AB, Palmer C, Mossman BT, Taatjes DJ [2003]. Persistent localization of activated extracellular signal-regulated kinases (ERK1/2) is epithelial cell-specific in an inhalation model of asbestosis. Am J Pathol 162:713–720.

Dai YT, Yu CP [1988]. Alveolar deposition of fibers in rodents and humans. J Aerosol Med 11:247–258.

Davis JMG [1994]. The role of clearance and dissolution in determining the durability or biopersistence of mineral fibers. Environ Health Perspect 102:113–117.

Davis JM, Addison J, Bolton RE, Donaldson K, Jones AD [1986]. Inhalation and injection studies in rats using dust samples from chrysotile asbestos prepared by a wet dispersion process. Br J Exp Pathol 67:113–129.

Davis JM, Addison J, McIntosh C, Miller BG, Niven K [1991]. Variations in the carcinogenicity of tremolite dust samples of differing morphology. Ann NY Acad Sci 643:473–490.

Davis LK, Martin TR, Kligler B [1992]. Use of death certificates for mesothelioma surveillance. Public Health Rep 107:481–483.

Davis JMG, Brown DM, Cullen RT, Donaldson K, Jones AD, Miller BC, Mcintosh C, Searl A [1996]. A comparison of methods of determining and predicting the pathogenicity of mineral fibers. Inhal Toxicol 8:747–770.

Dement JM, Wallingford KE [1990]. Comparison of phase contrast and electron microscopic methods for evaluation of occupational asbestos exposures. Appl Occup Environ Hyg 5:242–247.

Dement JM, Myers D, Loomis D, Richardson D, Wolf S [2009]. Estimates of historical exposures by phase contrast and transmission electron microscopy in North Carolina USA asbestos textile plants. Appl Occup Environ Hyg 66:574–583.

Dement JM, Brown DP, Okun A [1994]. Follow-up study of chrysotile asbestos textile workers: cohort mortality and case-control analyses. Am J Ind Med 26:431–447.

Dement JM, Kuempel E, Zumwalde R, Smith R, Stayner L, Loomis D [2008]. Development of a fibre size-specific job-exposure matrix for airborne asbestos fibres. Occup Environ Med 65:605–612.

Dement JM, Zumwalde RD, Wallingford KM [1976]. Discussion paper: asbestos fiber exposures in a hard rock gold mine. Ann NY Acad Sci 271:345–352.

De Vuyst P, Karjalainen A, Dumortier P, Pairon J-C, Monsó E, Brochard P, Teschler H, Tossavainen A, Gibbs A [1998]. Guidelines for mineral fibre analyses in biological samples: report of the ERS Working Group. Eur Respir J 11:1416–1426.

De Vuyst P, Gevenois PA [2002]. Asbestosis. In: Hendrick DJ, Burge PS, Beckett WS, Churg A, eds. Occupational disorders of the lung: recognition, management, and prevention. Oxford, UK: WB Saunders, pp. 143–162.

Deye GJ, Gao P, Baron PA, Fernback J [1999]. Performance evaluation of a fiber length classifier. Aerosol Sci Technol 30:420–437.

Ding M, Dong Z, Chen F, Pack D, Ma WY, Ye J, Shi X, Castranova V, Vallyathan V [1999]. Asbestos induces activator protein-1 transactivation in transgenic mice. Cancer Res 59:1884–1889.

Dodson RF, Atkinson MAL, Levin JL [2003]. Asbestos fiber length as related to potential pathogenicity: a critical review. Am J Ind Med 44:291–297.

Dogan AU, Baris YI, Dogan M, Emri S, Steele I, Elmishad AG, Carbone M [2006]. Genetic predisposition to fiber carcinogenesis causes a mesothelioma epidemic in Turkey. Cancer Res 66:5063–5068.

Dogan AU, Dogan M [2008]. Re-evaluation and re-classification of erionite series minerals. Environ Geochem Health 4:355–366.

Donaldson K, Tran CL [2002]. Inflammation caused by particles and fibers. Inhal Toxicol 14:5–27.

Dorling M, Zussman J [1987]. Characteristics of asbestiform and non-asbestiform calcic amphiboles. Lithos 20:469–489.

Driscoll KE, Carter JM, Borm PJA [2002]. Antioxidant defense mechanisms and the toxicity of fibrous and nonfibrous particles. Inhal Toxicol 14:101–118.

Driscoll KE, Carter JM, Howard BW, Hassenbein D, Janssen YM, Mossman BT [1998]. Crocidolite activates NF-κB and MIP-2 gene expression in rat alveolar epithelial cells: role of mitochondrial-derived oxidants. Environ Health Perspect 106(Suppl 5):1171–1174.

Drumm K, Messner C, Kienast K [1999]. Reactive oxygen intermediate-release of fibre-exposed monocytes increases inflammatory cytokine-mRNA level, protein tyrosine kinase and NF-κB activity in co-cultured bronchial epithelial cells (BEAS-2B). Eur J Med Res 4:257–263.

Egerton RF [2005]. Physical principles of electron microscopy: an introduction to TEM, SEM, and AEM. New York: Springer, pp. 202.

Enterline PE, Henderson VL [1987]. Geographic patterns for pleural mesothelioma deaths in the United States, 1968–81. J Natl Cancer Inst 79:31–37.

EPA [1986]. Airborne asbestos health assessment update. Washington, DC: U.S. Environmental Protection Agency, Office of Health and Environment Assessment. Report No. EPA/600/8-84/003F.

EPA [1987]. Asbestos-containing materials in schools: final rule and notice. 40 CFR Part 763. Fed Reg 52:41826–41905.

EPA [2001]. Federal Insecticide, Fungicide, and Rodenticide Act (FIFRA): FIFRA Scientific Advisory Panel Meeting, September 26, 2000. Test guidelines for chronic inhalation toxicity and carcinogenicity of fibrous particles. SAP Report No. 2001-01, January 5 [http://www.epa.gov/scipoly/sap/meetings/2000/september/final_fibers.pdf]. Date accessed: November 16, 2006.

EPA [2001]. OPPTS 870.8355: combined chronic toxicity/carcinogenicity testing of respirable fibrous particles. Report No. EPA 712–C01–352 [http://www.epa.gov/opptsfrs/home/guidelin.htm]. Date accessed: January 15, 2010.

EPA [2003]. Report on the Peer Consultation Workshop to discuss a proposed protocol to assess asbestos-related risk: final report. Washington, DC: Office of Solid Waste and Emergency Response, p. viii [http://www.epa.gov/oswer/riskassessment/asbestos/pdfs/asbestos_report.pdf]. Date accessed: June 30, 2008.

EPA [2008a]. SAB Consultation on EPA's Proposed Approach for estimation of bin-specific cancer potency factors for inhalation exposure to asbestos. [http://nepis.epa.gov/Exe/ZyNET.exe/P1002EAG.TXT?ZyActionD=ZyDocument&Client=EPA&Index=2006+Thru+2010&Docs=&Query=SAB09004%20or%20SAB%20or%20bin%20or%20specific%20or%20asbestos&Time=&EndTime=&SearchMethod=1&TocRestrict=n&Toc=&TocEntry=&QField=pubnumber%5E%22SAB09004%22&QFieldYear=&QFieldMonth=&QFieldDay=&UseQField=pubnumber&IntQFieldOp=1&ExtQFieldOp=1&XmlQuery=&File=D%3A%5Czyfiles%5CIndex%20Data%5C06thru10%5CTxt%5C00000005%5CP1002EAG.txt&User=ANONYMOUS&Password=anonymous&SortMethod=h%7C-&MaximumDocuments=10&FuzzyDegree=0&ImageQuality=r75g8/r75g8/x150y150g16/i425&Display=p%7Cf&DefSeekPage=x&SearchBack=ZyActionL&Back=ZyActionS&BackDesc=Results%20page&MaximumPages=1&ZyEntry=1&SeekPage=x&ZyPURL]. Date accessed: April 13, 2011.

EPA [2008b]. Letter from EPA Administrator to the Asbestos Committee Chair, December 29, 2008 [http://yosemite.epa.gov/sab/sabproduct.nsf/77CFF6439C00ABF3852575010077801F/$File/EPA-SAB-09-004+Response+12-29-2008.pdf]. Date accessed: April 13, 2011.

European Commission [1997]. Commission directive 97/69/EC of 5.XII.97 (23 adaptation) O.J. L 343/1997. Brussels: European Commission.

European Commission [1999]. Sub-chronic inhalation toxicity of synthetic mineral fibres in rats (ECB/TM/16 (97) Rev 1). In: Bernstein DM, Riego-Sintes JM, eds. Methods for the determination of the hazardous properties for human health of manmade mineral fibres (MMMF). Report EUR 18748 EN (1999). Brussels: European Commission Joint Research Centre [http://ecb.jrc.it/testing-methods]. Date accessed: November 16, 2006.

European Parliament and Council [2003]. Directive 2003/18/EC of the European Parliament and of the Council of 27 March 2003 amending Council Directive 83/477/EEC on the protection of workers from the risks related to exposure to asbestos at work. Off J Euro Union 15 April(L97):48–52.

Faux SP, Houghton CE, Hubbard A, Patrick G [2000]. Increased expression of epidermal growth factor receptor in rat pleural mesothelial cells correlates with carcinogenicity of mineral fibres. Carcinogenesis 21:2275–2280.

Faux SP, Howden PJ, Levy LS [1994]. Iron-dependent formation of 8-hydroxydeoxyguanosine in isolated DNA and mutagenicity in Salmonella typhimurium TA102 induced by crocidolite. Carcinogenesis 15:1749–1751.

Frank AL, Dodson RF, Williams MG [1998]. Carcinogenic implications of the lack of tremolite in UICC reference chrysotile. Am J Ind Med 34:314–317.

Franzblau A, Kazerooni EA, Sen A, Goodsitt M, Lee S-Y, Rosenman K, Lockey J, Meyer C, Gillespie B, Wang ML, Petsonk EL [2009]. Comparison of digital radiographs with film radiographs for the classification of pneumoconiosis. Acad Radiol 16:669–677.

Friesen MC, Davies HW, Tescheke K, Ostry AS, Hertzman C, Demers PA [2007]. Impact of the specificity of the exposure metric on exposure-response relationships. Epidemiology 18:88–94.

Fubini B [1993]. The possible role of surface chemistry in the toxicity of inhaled fibers. In: Warheit DB, ed., Fiber toxicology. Boston: Academic Press, pp. 229–257.

Gamble JF [1993]. A nested case control study of lung cancer among New York talc workers. Int Arch Occup Environ Health 64:449–456.

Gamble JF, Greife A, Hancock J [1982]. An epidemiological-industrial hygiene study of talc workers. Ann Occup Hyg 26:841–859.

Gamble JF, Gibbs GW [2008]. An evaluation of the risks of lung cancer and mesothelioma from exposure to amphibole cleavage fragments. Regul Toxicol Pharmacol 52:S154–S186.

Gendek EG, Brody AR [1990]. Changes in lipid ordering of model phospholipid membranes treated with chrysotile and crocidolite asbestos. Environ Res 53:152–167.

Gibbs GW, Hwang CY [1980]. Dimension of airborne asbestos fibres. In: Wagner JC, ed. Biological effects of mineral fibers. IARC Scientific Publications No. 30. Lyon, France: IARC, pp. 69–77.

Gillam J, Dement J, Lemen R, Wagoner J, Archer V, Blejer H [1976]. Mortality patterns among hard rock gold miners exposed to an asbestiform mineral. Ann NY Acad Sci 271:336–344.

Gilmour PS, Beswick PH, Brown DM, Donaldson K [1995]. Detection of surface free radical activity of respirable industrial fibers using supercoiled phi X174 plasmid DNA. Carcinogenesis 16:2973–2979.

Goldstein J [2003]. Scanning electron microscopy and x-ray microanalysis. New York: Kluwer Academic/Plenum Publishers, pp. 689

Goodglick LA, Kane AB [1986]. Role of reactive oxygen metabolites in crocidolite asbestos toxicity to mouse macrophages. Cancer Res 46:5558–5566.

Green GM [1973]. Alveolobronchiolar transport mechanisms. Arch Intern Med 131:109–114.

Green FHY, Harley R, Vallyathan V, Althouse R, Fick G, Dement J, Mitha R, Pooley F [1997]. Exposure and mineralogical correlates of pulmonary fibrosis in chrysotile asbestos workers. Occup Environ Med 54:549–559.

Greim HA [2004]. Research needs to improve risk assessment of fiber toxicity. Mut Res 553: 11–22.

Griffis LC, Pickrell JA, Carpenter RL, Wolff RK, McAllen SJ, Yerkes KL [1983]. Deposition of crocidolite asbestos and glass microfibers inhaled by the beagle dog. Am Ind Hyg Assoc J 44:216–222.

Gross P, DeTreville RT, Tolker EB, Kaschak M, Babyak MA [1967]. Experimental asbestosis: the development of lung cancer in rats with pulmonary deposits of chrysotile asbestos dust. Arch Environ Health 15:343–355.

Guthrie GD [1997]. Mineral properties and their contributions to particle toxicity. Environ Health Perspect 105(Suppl 5):1003–1011.

Hansen K, Mossman B [1987]. Generation of superoxide (O_2-•) from alveolar macrophages exposed to asbestiform and nonfibrous particles. Cancer Res 47:1681–1686.

Harper M, Bartolucci A [2003]. Preparation and examination of proposed consensus reference standards for fiber-counting. AIHA J 64: 283–287.

Harper M, Lee EG, Harvey B, Beard M [2007]. The effect of a proposed change to fiber-counting rules in ASTM International Standard D7200–06. J Occup Environ Hyg 4:D42–D45.

Harper M, Slaven JE, Pang TWS [2009]. Continued participation in an asbestos fiber-counting proficiency test with relocatable grid slides. J Environ Monit 11:434–438.

Harper M, Lee EG, Doorn SS, Hammond O [2008]. Differentiating non-asbestiform amphibole and amphibole asbestos by size characteristics. J Occup Environ Hyg 5:761–770.

HEI [1991]. Asbestos in public and commercial buildings: a literature review and synthesis of current knowledge. Cambridge, MA: Health Effects Institute [http://www.asbestos-institute.ca/reviews/hei-ar/hei-ar.html]. Date accessed: June 30, 2008.

Hein M, Stayner LT, Lehman E, Dement J [2007]. Follow-up study of chrysotile textile workers: cohort mortality and exposure response. Occup Environ Med 64:616–625.

Henderson DW, Jones ML, deKlerk N, Leigh J, Musk AW, Shilkin KB, Williams VM [2004]. The diagnosis and attribution of asbestos-related diseases in an Australian context. Int J Occup Environ Health 10:40–46.

Hesterberg TW, Barrett JC [1984]. Dependence of asbestos- and mineral dust-induced transformation of mammalian cells in culture on fiber dimension. Cancer Res 44:2170–2180.

Hesterberg TW, Hart GA [2000]. Lung biopersistence and in vitro dissolution rate predict the pathogenic potential of synthetic vitreous fibers. Inhal Toxicol 31:91–97.

HHS [2005a]. National Toxicology Program report on carcinogens, 11th ed. Washington, DC: U.S. Department of Health and Human Services [http://ntp.niehs.nih.gov/ntp/roc/eleventh/profiles/s016asbe.pdf]. Date accessed: December 28, 2006.

HHS [2005b]. National Toxicology Program report on carcinogens, 11th ed. Washington, DC: U.S. Department of Health and Human Services [http://ntp.niehs.nih.gov/ntp/roc/eleventh/profiles/s083erio.pdf]. Date accessed: November 16, 2006.

Higgins ITT, Glassman JH, Oh MS, Cornell RG [1983]. Mortality of reserve mining company employees in relation to taconite dust exposure. Am J Epidemiol 118:710–719.

Hill GD, Mangum JB, Moss OR, Everitt JI [2003]. Soluble ICAM-1, MCP-1, and MIP-2 protein secretion by rat pleural mesothelial cells following exposure to amosite asbestos. Exp Lung Res 29:277–290.

Hill IM, Beswick PH, Donaldson K [1996]. Enhancement of the macrophage oxidative burst by immunoglobulin coating of respirable fibers: fiber-specific differences between asbestos and man-made fibers. Exp Lung Res 22:133–148.

Hodgson JT, Darnton H [2000]. The quantitative risks of mesothelioma and lung cancer in relation to asbestos exposure. Ann Occup Hyg 44:565–601.

Hodgson JT, Darnton A [2010]. Mesothelioma risk from chrysotile. Occ Env Med 67:42.

Hochella MF [1993]. Surface chemistry, structure, and reactivity of hazardous mineral dust. In: Guthrie GD, Mossman BT, eds. Health effects of mineral dusts. Reviews in mineralogy, vol. 28. Washington, DC: Mineralogical Society of America, pp. 275–308.

Holmes S [1965]. Developments in dust sampling and counting techniques in the asbestos industry. Ann N Y Acad Sci 132:288–297.

Honda Y, Beall C, Delzell E, Oestenstad K, Brill I, Matthews R [2002]. Mortality among workers at a talc mining and milling facility. Ann Occup Hyg 46:575–585.

HSE (Health and Safety Executive) [1995]. Asbestos fibres in air: sampling and evaluation by phase contrast microscopy (PCM) under the Control of Asbestos at Work regulations (MDHS 39/4). Sudbury, MA: HSE Books.

Hull MJ, Abraham JL, Case BW [2002]. Mesothelioma among workers in asbestiform fiver-bearing talc mines in New York State. Ann Occup Hyg 46:132–132.

Hume LA, Rimstidt JD [1992]. The biodurability of chrysotile asbestos. Am Mineral 77: 1125–1128.

Huuskonen O, Kivisaari L, Zitting A, Taskinen K, Tossavainen A, Vehmas T [2001]. High-resolution computed tomography classification of lung fibrosis for patients with asbestos-related disease. Scand J Work Environ Health 27: 106–112.

Hwang C-Y, Gibbs GW [1981]. The dimensions of airborne asbestos fibres. I: crocidolite from Kuruman area, Cape Province, South Africa. Ann Occup Hyg 24:23–41.

Iakhiaev A, Pendurthi U, Idell S [2004]. Asbestos induces tissue factor in Beas-2B human lung bronchial epithelial cells *in vitro*. Lung 182:251–264.

IARC [1977]. IARC monographs on the evaluation of carcinogenic risk of chemicals to man. Vol. 14: asbestos. Lyon, France: International Agency for Research on Cancer, pp. 11–106.

IARC [1987a]. IARC monographs on the evaluation of carcinogenic risk of chemicals to humans. Vol. 42: silica and some silicates. Lyon, France: International Agency for Research on Cancer, pp. 33–249.

IARC [1987b]. IARC monographs on the evaluation of carcinogenic risks to humans. Suppl. 7: overall evaluations of carcinogenicity: an updating of IARC monographs. Vols. 1–42. Lyon,

France: International Agency for Research on Cancer, pp. 106–117, 203.

IARC [1997]. IARC summaries & evaluations: zeolites other than erionite (group 3). Lyon, France: International Agency for Research on Cancer [www.inchem.org/documents/iarc/vol68/zeol.html]. Date accessed: December 21, 2009.

IARC [2002]. IARC monographs on the evaluation of carcinogenic risks to humans: man-made vitreous fibers. Vol. 81. Lyon, France: International Agency for Research on Cancer [http://monographs.iarc.fr/ENG/Monographs/vol81/mono81.pdf]. Date accessed: June 30, 2008.

IARC [2009]. Special report: policy. A review of human carcinogens, part C: metals, arsenic, dusts, and fibres. Lancet *10*:453–454.

ICRP (International Commission on Radiation Protection) [1994]. Human respiratory tract model for radiological protection. ICRP publication 66(1994). Ann ICRP (UK) *24*(1–3).

Ilgren EB, Browne K [1991]. Asbestos-related mesothelioma: evidence for a threshold in animals and humans. Regul Toxicol Pharmacol *13*:116–132.

ILO [2002]. Guidelines for the use of the ILO International Classification of Pneumoconioses. Rev. ed. Occupational Safety and Health Series, No. 22. Geneva: International Labour Office.

ILSI (International Life Sciences Institute) [2005]. Testing of fibrous particles: short-term assays and strategies. Report of an ILSI Risk Science Institute Working Group. Inhal Toxicol *17*:497–537.

IMA-NA (Industrial Minerals Association–North America) [2005]. Submission to MSHA RIN 1219-AB24—Proposed rule—asbestos exposure limit. November 21, 2005 [http://www.msha.gov/regs/comments/05-14510/1219-ab24-comm-107.pdf]. Date accessed: June 30, 2008.

IOM (Institute of Medicine of the National Academies) Committee on Asbestos—Selected Health Effects [2006]. Asbestos: selected cancers. Washington, DC: The National Academies Press [http://books.nap.edu/openbook.php?record_id=11665&page=R1]. Date accessed: June 30, 2008.

ISO [1993]. Air quality: determination of the number concentration of airborne inorganic fibres by phase contrast optical microscopy — membrane filter method. ISO 8672. Geneva: International Organization for Standardization.

ISO [1995a]. Air quality: particle size fraction definitions for health-related sampling. ISO 7708. Geneva: International Organization for Standardization.

ISO [1995b]. Ambient air: determination of asbestos fibres—direct-transfer transmission electron microscopy method. ISO 10312. Geneva: International Organization for Standardization.

ISO [1999]. Ambient air: determination of asbestos fibres—indirect-transfer transmission electron microscopy method. ISO 13794. Geneva: International Organization for Standardization.

ISO [2002]. Ambient air: determination of numerical concentration of inorganic fibrous particles—scanning electron microscopy method. ISO 14966. Geneva: International Organization for Standardization.

Iwagaki A, Choe N, Li Y, Hemenway DR, Kagan E [2003]. Asbestos inhalation induces tyrosine nitration associated with extracellular signal-regulated kinase 1/2 activation in the rat lung. Am J Respir Cell Mol Biol *28*:51–60.

Janssen Y, Heintz N, Marsh J, Borm P, Mossman B [1994]. Induction of c-fos and c-jun proto-oncogenes in target cells of the lung and pleura by carcinogenic fibers. Am J Respir Cell Mol Biol *11*:522–530.

Järvholm B, Englund A, Albin M [1999]. Pleural mesothelioma in Sweden: an analysis of the incidence according to the use of asbestos. Occup Environ Med *56*:110–113.

Jaurand MC [1991]. Mechanisms of fiber genotoxicity. In: Brown RC, Hoskins JA, Johnson NF, eds. Mechanisms in fiber carcinogenesis. New York: Plenum, pp. 287–307.

Jaurand MC [1997]. Mechanisms of fiber-induced genotoxicity. Environ Health Perspect *105*(Suppl 5):1073–1084.

Jaurand MC, Baillif P, Thomassin JH, Magne L, Touray JC [1983]. X-ray photoelectron spectroscopy and chemical study of the adsorption of biological molecules on the chrysotile asbestos surface. J Colloid Interface Sci *95*:1–9.

Jaurand MC, Thomassin JH, Baillif P, Magne L, Touray JC, Bignon J [1980]. Chemical and photoelectron spectrometry analysis of the adsorption of phospholipid model membranes and red blood cell membranes on to chrysotile fibres. Br J Ind Med *37*:169–174.

Jones AD, Aitken RJ, Fabriès JF, Kauffer E, Lidén G, Maynard A, Riediger G, Sahle W [2005]. Thoracic size-selective sampling of fibres: performance of four types of thoracic sampler in laboratory tests. Ann Occup Hyg *49*:481–492.

Jones HA, Hamacher K, Hill AA, Clark JC, Krausz T, Boobis AR, Haslett C [1997]. 18-F fluoroproline (18FP) uptake monitored *in vivo* in a rabbit model of pulmonary fibrosis [abstract]. Am J Respir Crit Care Med *155*:A185.

Jurinski JB, Rimstidt JD [2001]. Biodurability of talc. Am Mineral *86*:392–399.

Kane AB [1991]. Fiber dimensions and mesothelioma: a reappraisal of the Stanton hypothesis. In: Brown RC, Hoskins, JA Johnson NF, eds. Mechanisms in fibre carcinogenesis. New York: Plenum, pp. 131–141.

Kane AB [1996]. Mechanisms of mineral fibre carcinogenesis. In: Kane AB, Saracci R, Weilbourn JD, eds. Mechanisms of fibre carcinogenesis. IARC Science Publications No. 140. Lyon, France: International Agency for Research on Cancer, pp. 11–34.

Karakoca Y, Emri S, Cangur AK, Baris YI [1997]. Environmental pleural plaques due to asbestos and fibrous zeolite exposure in Turkey. Indoor Built Environ *6*:100–105.

Keane MJ, Stephens JW, Zhong BZ, Miller WE, Wallace WE [1999]. A study of the effect of chrysotile fiber surface composition on genotoxicity *in vitro*. J Toxicol Environ Health *57*:529–541.

Kelse JW [2005]. White paper: asbestos, health risk and tremolitic talc. Norwalk, CT: RT Vanderbilt Co., Inc.

Kelse JW [2008]. Letter communication to the NIOSH Docket Office for Docket No. NIOSH–099C. [http://www.cdc.gov/niosh/docket/archive/pdfs/NIOSH-099A/0099A-092308-Kelse_sub.pdf]. Date accessed: September 29, 2010.

Kenny LC, Rood AP [1987]. A direct measurement of the visibility of amosite asbestos fibres by phase contrast optical microscopy. Ann Occup Hyg *31*:261–264.

Kleinfeld M, Messite J, Kooyman O, Zaki M [1967]. Mortality among talc miners and millers in New York State. Arch Environ Health 14:663–667.

Kleinfeld M, Messite J, Tabershaw IR [1955]. Talc pneumoconiosis. AMA Arch Ind Health 12:66–72.

Kleinfeld M, Messite J, Zacki MH [1974]. Mortality experiences among talc workers: a follow-up study. J Occup Med 16:345–349.

Kliment CR, Clemens K, Oury TD [2009]. North American erionite-associated mesothelioma with pleural plaques and pulmonary fibrosis: a case report. Int J Clin Exp Pathol 2:407–410.

Kohyama N, Shinohara Y, Suzuky Y [1996]. Mineral phases and some reexamined characteristics of the Internation Union Against Cancer standard asbestos samples. Am J Ind Med 30:515–528.

Kraus T, Raithel HJ, Hering KG, Lehnert G [1996]. Evaluation and classification of high-resolution computed tomographic findings in patients with pneumoconiosis. Int Arch Occup Environ Health 68:249–254.

Kuempel ED, O'Flaherty EJ, Stayner LT, Smith RJ, Green FHY, Vallyathan V [2001]. A biomathematical model of particle clearance and retention in the lungs of coal miners. Part I: model development. Regul Toxicol Pharmacol 34:69–87.

Kuempel ED, Stayner LT, Dement JD, Gilbert SJ, Hein MJ [2006]. Fiber size-specific exposure estimates and updated mortality analysis of chrysotile asbestos textile workers [abstract #349]. Toxicol Sci 90:71.

Lamm SH, Starr JA [1988]. Similarities in lung cancer and respiratory disease mortality of Vermont and New York State talc workers. In: Proceedings of the 7th International Pneumoconioses Conference. Cincinnati, OH: U.S. Department of Health and Human Services, Centers for Disease Control, National Institute for Occupational Safety and Health. DHHS (NIOSH) Publication No. 90-108, Part II, pp. 1576–1581.

Lamm SH, Levine MS, Starr JA, Tirey SL [1988]. Analysis of excess lung cancer risk in short-term employees. Am J Epidemiol 127: 1202–1209.

Langer AM, Nolan RP, Addison J [1991]. Distinguishing between amphibole asbestos fibers and elongate cleavage fragments of their non-asbestos analogues. In: Brown RC, Hoskins JA, Johnson NF, eds. Mechanisms in fibre carcinogenesis. New York: Plenum Press, pp. 231–251.

Larsen ES [1942]. Alkalic rocks of Iron Hill, Gunnison County, Colorado. U.S. Geological Survey Professional Paper 197–A. Reston, VA: U.S. Geological Survey, 64 pp.

Lawler AB, Mandel JS, Schuman LM, Lubin JH [1985]. A retrospective cohort mortality study of iron ore (hematite) miners in Minnesota. J Occup Med 27:507–517.

Leake BE [1978]. Nomenclature of amphiboles. Can Mineral 16:501–520.

Leake BE, Woolley AR, Arps CES, Birch WD, Gilbert CM, Grice JD, Hawthorne FC, Kato A, Kisch HF, Krivovichev VG, Linthout K, Laird J, Mandarino JA, Maresch WV, Nickel EH, Rock NMS, Schumacher JC, Smith DC, Stephenson NCN, Ungaretti L, Whittaker EJW, Youzhi G [1997]. Nomenclature of the amphiboles: report of the Subcommittee on Amphiboles of the International Mineralogical Association Commission on new minerals and mineral names. Can Mineral 35:219–246.

Lee EG, Harper M, Nelson J, Hintz PJ, Andrew ME [2008]. A comparison of the CATHIA-T sampler, the GK2.69 cyclone and the standard cowled sampler for thoracic fiber concentrations at a taconite ore-processing mill. Ann Occup Hyg 52:55–62.

Lee EG, Nelson J, Hintz PJ, Joy G, Andrew ME, Harper M [2010]. Field performance of the CATHIA-T sampler and two cyclones against the standard cowled sampler for thoracic fiber concentrations. Ann Occup Hyg 54:545–556.

Lee YCG, deKlerk NH, Henderson DK, Musk AW [2002]. Malignant mesothelioma. In: Hendrick DJ, Burge PS, Beckett WS, Churg A, eds. Occupational disorders of the lung: recognition, management, and prevention. Oxford, UK: WB Saunders, pp. 359–379.

Leineweber JP [1984]. Solubility of fibers *in vitro* and *in vivo*. In: Biological effects of man-made mineral fibers. Vol. 2. Copenhagen: World Health Organization, pp. 87–101.

Light WG, Wei ET [1977a]. Surface charge and hemolytic activity of asbestos. Environ Res 13:135–145.

Light WG, Wei ET [1977b]. Surface charge and asbestos toxicity. Nature 265:537–539.

Lilienfeld DE, Mandel JS, Coin P, Schuman LM [1988]. Projection of asbestos related diseases in the United States, 1985–2009. I: cancer. Br J Ind Med 45:283–291.

Lippmann M [1988]. Asbestos exposure indices. Environ Res 46:86–106.

Lippmann M [1990]. Effects of fiber characteristics on lung deposition, retention, and disease. Environ Health Perspect 88:311–317.

Lippmann M, Esch JL [1988]. Effect of lung airway branching pattern and gas composition on particle deposition. I: background and literature review. Exp Lung Res 14:311–320.

Lippmann M, Schlesinger RB [1984]. Interspecies comparisons of particle deposition and mucociliary clearance in tracheobronchial airways. J Toxicol Environ Health 14:141–169.

Lippmann M, Yeates, Albert RE [1980]. Deposition, retention, and clearance of inhaled particles. Br J Ind Med 37:337–362.

Loomis D, Dement JM, Wolf SH, Richardson DB [2009]. Lung cancer mortality and fibre exposures among North Carolina asbestos textile workers. Occ Env Med 66:535–542.

Lowers H, Meeker G [2002]. Tabulation of asbestos-related terminology. Open-file report 02–458, p. 70, [http://pubs.usgs.gov/of/2002/ofr-02-458/]. Date accessed: December 21, 2009.

Lu J, Keane MJ, Ong T, Wallace WE [1994]. *In vitro* genotoxicity studies of chrysotile asbestos fibers dispersed in simulated pulmonary surfactant. Mutat Res 320:253–259.

Luce D, Bugel I, Goldberg P, Goldberg M, Salomon C, Billon-Galland MA, Nicolau J, Quénel P, Fevotte J, Brochard P [2000]. Environmental exposure to tremolite and respiratory cancer in New Caledonia: a case-control study. Am J Epidemiol 151:259–265.

Lynch JR, Ayer HE, Johnson DL [1970]. The interrelationships of selected asbestos exposure indices. Am Ind Hyg Assoc J 31:598–604.

Maimon M [2010]. Letter communication to the NIOSH Docket Office for Docket No. NIOSH–099C [http://www.cdc.gov/niosh/docket/archive/pdfs/NIOSH-099C/0099C-040910-Maimon_M_sub.pdf]. Date accessed: August 19, 2010.

Mandel J (mand0125@umn.edu) [2008]. Question about Conwed. Private e-mail message to Paul Middendorf (pkm2@cdc.gov), February 27.

Maples KR, Johnson NF [1992]. Fiber-induced hydroxyl radical formation: correlation with mesothelioma induction in rats and humans. Carcinogenesis 13:2035–2039.

Marchant GE, Amen MA, Bullock CH, Carter CM, Johnson KA, Reynolds JW, Connelly FR, Crane AE [2002]. A synthetic vitreous fiber (SVF) occupational exposure database: implementing the SVF health and safety partnership program. App Occup Environ Hyg 17:276–285.

Marconi A, Menichini E, Paolleti L [1984]. A comparison of light microscopy and transmission electron microscopy results in the evaluation of the occupational exposure to airborne chrysotile fibres. Ann Occup Hyg 28:321–331.

Marsh JP, Mossman BT [1988]. Mechanisms of induction of ornithine decarboxylase activity in tracheal epithelial cells by asbestiform minerals. Cancer Res 48:709–714.

Mast RW, Maxim LD, Utell MJ, Walker AM [2000]. Refractory ceramic fiber: toxicology, epidemiology, and risk analyses—a review. Inhal Toxicol 12:359–399.

Maxim LD, McConnell EE [2001]. Interspecies comparisons of the toxicity of asbestos and synthetic vitreous fibers: a weight-of-the-evidence approach. Regul Toxicol Pharmacol 33:319–342.

Maynard A [2002]. Thoracic size–selection of fibres: dependence of penetration on fibre length for five thoracic samplers. Ann Occup Hyg 46:511–522.

Maynard A, Aitken RJ, Butz T, Colvin V, Donaldson, K, Oberdorster G, Philbert MA, Ryan J, Seaton A, Stone V, Tinkle SS, Tran L, Walker NG, Warheit D 2006]. Safe handling of nanotechnology. Nature 444:267–269.

McDonald JC, McDonald AD [1997]. Chrysotile, tremolite and carcinogenicity. Ann Occup Hyg 41:699–705.

McDonald AD, Fry JS, Woolley AJ [1983]. Dust exposure and mortality in an American chrysotile textile plant. Br J Ind Med 40:361–367.

McDonald JC, Gibbs GW, Liddel FDK, McDonald AD [1978]. Mortality after long exposure to cummingtonite-grunerite. Am Rev Respir Dis 118:271–277.

McDonald JC, Harris J, Armstrong B [2004]. Mortality in a cohort of vermiculite miners exposed to fibrous amphibole in Libby, Montana. Occup Environ Med 61:363–366.

Minnesota Department of Health [2007]. Mesothelioma in Northeastern Minnesota and two occupational cohorts: 2007 update. St. Paul, MN: Center for Occupational Health and Safety, Chronic Disease and Environmental Epidemiology Section, Minnesota Department of Health [http://www.health.state.mn.us/divs/hpcd/cdee/mcss/documents/nemeso1207.pdf]. Date accessed: June 30, 2008.

Meeker GP, Bern AM, Brownfield IK, Lowers HA, Sutley SJ, Hoefen TM, Vance JS [2003]. The composition and morphology of amphiboles from the Rainy Creek Complex, near Libby, Montana. Am Mineral 88:1955–1969.

Menvielle G, Luce D, Févotte J, Bugel I, Salomon C, Goldberg P, Billon-Galland MA, Goldberg M [2003]. Occupational exposures and lung cancer in New Caledonia. Occup Environ Med 60:584–589.

Mesothelioma Virtual Bank [2007]. Mesothelioma tissue resources available for your research (Web site update: November 29, 2007) [http://www.mesotissue.org/]. Date accessed: January 10, 2007.

Middendorf P, Graff R, Keller L, Simmons C [2007]. National Occupational Exposure Database: AIHA-NIOSH alliance efforts to develop a pilot. Presented at American Industrial Hygiene Conference and Exhibition (AIHce), Philadelphia, PA.

Miller A [2007]. Radiographic readings for asbestosis: misuse of science—validation of the ILO classification. Am J Ind Med 50:63–67.

Miller AL, Hoover, MD, Mitchell DM, Stapleton BP [2007]. The Nanoparticle Information Library (NIL): a prototype for linking and sharing emerging data. J Occup Environ Hyg 4:D131–D134.

Moolgavkar SH, Brown RC, Turim J [2001]. Biopersistence, fiber length, and cancer risk assessment for inhaled fibers. Inhal Toxicol 13:755–772.

Morrow PE [1985]. Pulmonary clearance. In: Hatch TF, Esmen NA, Mehlman MA, eds. Advances in modern environmental toxicology. Vol. 7. Occupational and Industrial Hygiene: Concepts and Methods. Princeton, NJ: Princeton Scientific Publishers, pp. 183–202.

Mossman BT [2008]. Assessment of the pathogenic potential of asbestiform vs. nonasbestiform particulates (cleavage fragments) in *in vitro* (cell or organ culture) models and bioassays. Regul Toxicol Pharmacol 52(Suppl 1): S200–S203.

Mossman BT, Marsh JP [1989]. Evidence supporting a role for active oxygen species in asbestos-induced toxicity and lung disease. Environ Health Perspect 81:91–94.

Mossman B, Sesko A [1990]. *In vitro* assays to predict the pathogenicity of mineral fibers. Toxicology 60:53–61.

Mossman BT, Borm PJ, Castranova V, Costa DL, Donaldson K, Kleeberger SR [2007]. Mechanisms of action of inhaled fibers, particles and nanoparticles in lung and cardiovascular diseases. Part Fibre Toxicol 4:4.

Mossman BT, Faux S, Janssen Y, Jimenez LA, Timblin C, Zanella C, Goldberg J, Walsh E, Barchowsky A, Driscoll K [1997]. Cell signaling pathways elicited by asbestos. Environ Health Perspect 105(Suppl 5):1121–1125.

Mossman BT, Jean L, Landesman JM [1983]. Studies using lectins to determine mineral interactions with cellular membranes. Environ Health Perspect 51:23–25.

Mossman BT, Lounsbury KM, Reddy SP [2006]. Oxidants and signaling by mitogen-activated protein kinases in lung epithelium. Am J Respir Cell Mol Biol 34:666–669.

MSHA [2002]. Mine employment and commodity data. Arlington, VA: U.S. Department of Labor, Mine Safety and Health Administration, Directorate of Program Evaluation and Information Resources [www.msha.gov/STATS/STATINFO.htm]. Date accessed: August 13, 2008.

Mine Safety and Health Administration [2005]. Asbestos exposure limit; proposed rule. Fed Reg. July 29:43950–43989 [http://edocket.access.gpo.gov/2005/pdf/05-14510.pdf]. Date accessed: June 30, 2008.

Mine Safety and Health Administration [2008]. Asbestos exposure limit; final rule. Fed Reg. February 29:11283–11304 [http://edocket.access.

gpo.gov/2008/pdf/E8-3828.pdf]. Date accessed: June 30, 2008.

Muhle H, Pott F [2000]. Asbestos as a reference material for fiber-induced cancer. Int Arch Occup Environ Health 73:53–59.

Muhle H, Pott F, Bellmann B, Takenaka S, Ziem U [1987]. Inhalation and injection experiments in rats for testing MMMF on carcinogenicity. Ann Occup Hyg 31:755–764.

Myojo T [1999]. A simple method to determine the length distribution of fibrous aerosols. Aerosol Sci Technol 30:30–39.

Nagle JF [1993]. Area/lipid of bilayers from NMR. Biophys J 64:1476–1481.

NIOSH [1976]. Revised recommended asbestos standard. Cincinnati, OH: U.S. Department of Health, Education, and Welfare, Center for Disease Control, National Institute for Occupational Safety and Health, DHEW (NIOSH) Publication No. 77–169 [http://www.cdc.gov/niosh/docs/77-169/]. Date accessed: June 30, 2008.

NIOSH [1980]. Occupational exposure to talc containing asbestos. Cincinnati, OH: U.S. Department of Health, Education, and Welfare, Center for Disease Control, National Institute for Occupational Safety and Health, DHEW (NIOSH) Publication No. 80–115 [http://www.cdc.gov/niosh/review/public/099/pdfs/TalcContainingAsbestosTR.pdf]. Date accessed: June 30, 2008.

NIOSH [1984]. NIOSH testimony to the U.S. Department of Labor: statement of the National Institute for Occupational Safety and Health. Presented at the public hearing on occupational exposure to asbestos, June 21, 1984. NIOSH policy statements. Cincinnati, OH: U.S. Department of Health and Human Services, Centers for Disease Control, National Institute for Occupational Safety and Health.

NIOSH [1990a]. Comments of the National Institute for Occupational Safety and Health on the Occupational Safety and Health Administration's Notice of Proposed Rulemaking on Occupational Exposure to Asbestos, Tremolite, Anthophyllite, and Actinolite. OSHA Docket No. H–033d, April 9, 1990 [http://www.cdc.gov/niosh/review/public/099/pdfs/AsbestosTestimony_April%209_1990.pdf]. Date accessed: June 30, 2008.

NIOSH [1990b]. Testimony of the National Institute for Occupational Safety and Health on the Occupational Safety and Health Administration's Notice of Proposed Rulemaking on Occupational Exposure to Asbestos, Tremolite, Anthophyllite, and Actinolite. OSHA Docket No. H–033d, May 9, 1990 [http://www.cdc.gov/niosh/review/public/099/pdfs/asbestos_testimony_May9.pdf]. Date accessed: June 30, 2008.

NIOSH [1994a]. Method 7400: asbestos and other fibers by PCM. Issue 2 (8/15/94). In: NIOSH manual of analytical methods. 4th ed. Cincinnati, OH: U.S. Department of Health and Human Services, Centers for Disease Control and Prevention, National Institute for Occupational Safety and Health, DHHS (NIOSH) Publication No. 2003–154 [http://www.cdc.gov/niosh/nmam/pdfs/7400.pdf]. Date accessed: June 30, 2008.

NIOSH [1994b]. Method 7402: asbestos by TEM. Issue 2 (8/15/94). In: NIOSH manual of analytical methods. 4th ed. Cincinnati, OH: U.S. Department of Health and Human Services, Centers for Disease Control and Prevention, National Institute for Occupational Safety and Health, DHHS (NIOSH) Publication No. 2003–154 [http://www.cdc.gov/niosh/nmam/pdfs/7402.pdf]. Date accessed: June 30, 2008.

NIOSH [1995]. Report to Congress on Workers' Home Contamination Study Conducted Under the Workers' Family Protection Act (29 U.S.C. 671a). Cincinnati, OH: U.S. Department of Health and Human Services, Centers for Disease Control and Prevention, National Institute for Occupational Safety and Health, DHHS (NIOSH) Publication No. 95–123 [http://www.cdc.gov/niosh/95-123.html]. Date accessed: September 9, 2010.

NIOSH [1998]. Criteria for a Recommended Standard: Occupational Exposure to Metalworking Fluids. U.S. Department of Health and Human Services, Centers for Disease Control and Prevention, Cincinnati, OH, DHHS (NIOSH) Publication No. 98–102 [http://www.cdc.gov/niosh/98-102.html]. Date accessed: March 14, 2008.

NIOSH [2002]. Comments of the National Institute for Occupational Safety and Health on the Mine Safety and Health Administration Advanced Notice of Proposed Rulemaking on Measuring and Controlling Asbestos Exposure, June 27, 2002 [http://www.msha.gov/regs/comments/asbestos/docket/comments/ab24comm-31.pdf]. Date accessed: June 29, 2008.

NIOSH [2003a]. Monitoring of diesel particulate exhaust in the workplace. Chapter Q. In: NIOSH manual of analytical methods. 4th ed. Cincinnati, OH: U.S. Department of Health and Human Services, Centers for Disease Control and Prevention, National Institute for Occupational Safety and Health, DHHS (NIOSH) Publication No. 2003–154 [http://www.cdc.gov/niosh/nmam/pdfs/chapter-q.pdf]. Date accessed: June 30, 2008.

NIOSH [2003b]. Method 7603: quartz in coal mine dust, by IR (redeposition). Issue 3 (3/15/03). In: NIOSH manual of analytical methods. 4th ed. Cincinnati, OH: U.S. Department of Health and Human Services, Centers for Disease Control and Prevention, National Institute for Occupational Safety and Health, DHHS (NIOSH) Publication No. 2003–154 [http://www.cdc.gov/niosh/nmam/pdfs/7603.pdf]. Date accessed: June 30, 2008.

NIOSH [2005]. Comments on the MSHA Proposed Rule on Asbestos Exposure Limit, October 13, 2005 [http://www.msha.gov/regs/comments/05-14510/1219-ab24-comm-103.pdf]. Date accessed: June 29, 2008.

NIOSH [2007a]. Asbestos: geometric mean exposures by major industry division, MSHA and OSHA samples, 1979–2003 [http://www2a.cdc.gov/drds/WorldReportData/FigureTableDetails.asp?FigureTableID=503&GroupRefNumber=F01-05]. Date accessed: June 29, 2008.

NIOSH [2007b]. National Occupational Respiratory Mortality System (NORMS) [http://webappa.cdc.gov/ords/norms.html]. Date accessed: January 26, 2007.

NIOSH [2007c]. Chest radiography B Reader information for medical professionals [http://www.cdc.gov/niosh/topics/chestradiography/breader-info.html]. Date accessed: April 15, 2008.

NIOSH [2007d]. Chest radiography: ethical considerations for B readers (Topic Page posted March 28, 2007) [http://www.cdc.gov/niosh/topics/chestradiography/breader-ethics.html]. Date accessed: April 15, 2008.

NIOSH [2007e]. Chest radiography: recommended practices for reliable classification of chest radiographs by B readers (Draft Topic Page posted March 28, 2007) [http://www.cdc.gov/niosh/topics/chestradiography/radiographic-classification.html]. Date accessed: January 9, 2008.

NIOSH [2008a]. Application of the ILO International Classification of Radiographs of

Pneumoconioses to Digital Chest Radiographic Images: A NIOSH Scientific Workshop. Cincinnati, OH: U.S. Department of Health and Human Services, Centers for Disease Control and Prevention, National Institute for Occupational Safety and Health, DHHS (NIOSH) Publication No. 2008–139 [http://www.cdc.gov/niosh/docs/2008-139/]. Date accessed: November 5, 2008.

NIOSH [2008b]. Strategic plan for NIOSH nanotechnology research and guidance: filling the knowledge gaps [http://www.cdc.gov/niosh/topics/nanotech/strat_plan.html]. Date accessed: June 11, 2008.

NIOSH [2009a]. Malignant mesothelioma: age-adjusted death rates by county, U.S. residents age 15 and over, 2000–2004 [http://www2a.cdc.gov/drds/WorldReportData/FigureTableDetails.asp?FigureTableID=900&GroupRefNumber=F07-03]. Date Accessed: August 23, 2010.

NIOSH [2009b]. Asbestosis: age-adjusted death rates by county, U.S. residents age 15 and over, 1995–2004. [www2a.cdc.gov/drds/WorldReportData/FigureTableDetails.asp?FigureTableID=500&GroupRefNumber=F01-03b]. Date accessed: August 23, 2010.

Nolan RP, Langer AM, Oechsle GE, Addison J, Colflesh DE [1991]. Association of tremolite habit with biological potential. In: Brown RC, Hoskins JA, Johnson NF, eds. Mechanisms in fibre carcinogenesis. New York: Plenum Press, pp. 231–251.

Nordmann M, Sorge A [1941]. Lungenkrebs durch asbestsaub im tierversuch [abstracted in IARC 1977]. Z Krebsforsch 51:168–182.

NRC (National Research Council) [1984]. Asbestiform fibers: nonoccupational health risks. Washington, DC: National Academy Press [http://www.nap.edu/openbook.php?isbn=0309034469]. Date accessed: June 30, 2008.

NSSGA (National Stone Sand and Gravel Association) [2005]. Submission to MSHA RIN 1219–AB24—Proposed Rule—Asbestos Exposure Limit, November 18, 2005 [http://www.msha.gov/regs/comments/05-14510/1219-ab24-comm-110.pdf]. Date accessed: June 30, 2008.

NTP (National Toxicology Program) [2005]. Report on Carcinogens, Eleventh Edition; U.S. Department of Health and Human Services, Public Health Service [http://ntp.niehs.nih.gov/ntp/roc/eleventh/profiles/s016asbe.pdf]. Date accessed: April 13, 2011.

Oberdorster G [1994]. Macrophage-associated responses to chrysotile. Ann Occup Hyg 38:601–615.

Oberdorster G, Morrow PE, Spurny K [1988]. Size dependent lymphatic short term clearance of amosite fibers in the lung. Ann Occup Hyg 32:149–156.

Oehlert GW [1991]. A reanalysis of the Stanton et al. pleural sarcoma data. Environ Res 54:194–205.

Oestenstad K, Honda Y, Delzell E, Brill I [2002]. Assessment of historical exposures to talc at a mining and milling facility. Ann Occup Hyg 46:587–596.

Okayasu R, Wu L, Hei TK [1999]. Biological effects of naturally occurring and man-made fibres: *in vitro* cytotoxicity and mutagenesis in mammalian cells. Br J Cancer 79:1319–1324.

Ollikainen T, Linnainmaa K, Kinnula VL [1999]. DNA single strand breaks induced by asbestos fibers in human pleural mesothelial cells *in vitro*. Environ Mol Mutagen 33:153–160.

OSHA [1986]. Asbestos; final rules. Fed Reg 51:22612–22790.

OSHA [1990]. Occupational exposure to asbestos, tremolite, anthophyllite and actinolite; proposed rulemaking (supplemental) and notice of hearing. Fed Reg 55:29712–29753.

OSHA [1992]. Occupational exposure to asbestos, tremolite, anthophyllite and actinolite. Preamble to Final Rule, Section 5–V. Health Effects. 57 Fed Regist 24310–24330, June 8, 1992.

OSHA [1998]. Sampling and analytical methods, asbestos in air, method ID-160 [http://www.osha.gov/dts/sltc/methods/inorganic/id160/id160.html]. Date accessed: April 8, 2008.

OSHA [2008]. Safety and health topics: asbestos [http://www.osha.gov/SLTC/asbestos/index.html]. Date accessed: January 28, 2008.

Pang TWS [2000]. Precision and accuracy of asbestos fiber counting by phase contrast microscopy. Am Ind Hyg Assoc J 61:529–538.

Pang TWS [2002]. The quality of fiber count data of slides with relocatable fields. Presented at the 2002 Johnson Conference: A Review of Asbestos Monitoring Methods and Results for the New York World Trade Center, Libby Vermiculite, and Fibrous Talc, July 21–25, 2002, Johnson State College, Johnson, VT.

Pang TWS, Harper M [2008]. The quality of fiber counts using improved slides with relocatable fields. J Environ Monit 10:89–95.

Pang TWS, Dicker WL, Nazar MA [1984]. An evaluation of the precision and accuracy of the direct transfer method for the analysis of asbestos fibers with comparison to the NIOSH method. Am Ind Hyg Assoc J 45:329–335.

Pang TWS, Schonfeld FA, Patel K [1989]. An improved membrane filter technique for evaluation of asbestos fibers. Am Ind Hyg Assoc J 50:174–180.

Pelé JP, Calvert R [1983]. Hemolysis by chrysotile asbestos fibers. I: influence of the sialic acid content in human, rat, and sheep red blood cell membranes. J Toxicol Environ Health 12:827–840.

Petersen EU, Totten F, Guida M [1993]. Tremolite-talc occurrences in the Balmat-Edwards District. In: Petersen EU, Slack J, eds. Selected mineral deposits of Vermont and the Adirondack Mountains, N.Y. Guidebook Series. Vol. 17. New York: Society of Economic Geologists, pp. 54–64.

Peto J, Seidman H, Selikoff IJ [1982]. Mesothelioma mortality in asbestos workers: implications for models of carcinogenesis and risk assessment. Br J Cancer 45:124–135.

Poland CA, Duffin R, Kinloch I, Maynard A, Wallace WAH, Seaton A, Stone V, Brown S, MacNee W, Donaldson K [2008]. Carbon nanotubes introduced into the abdominal cavity of mice show asbestos-like pathogenicity in a pilot study. Nat Nanotechnol 3:423–428.

Porter DW, Hubbs AF, Mercer RR, Wu N, Wolfarth MG, Sriram K, Leonard SS, Tatelli L, Schwegler-Berry D, Friend S, Andrew M, Chen Bt, Tsuruoka S, Endo M, Castranova V [2009]. Mouse pulmonary dose- and time course- responses induced by exposure to multi-walled carbon nanotubes. Toxicology 269:136–147.

Pott F, Huth F, Friedrichs KH [1974]. Tumorigenic effect of fibrous dusts in experimental animals. Environ Health Perspect 9:313–315.

Pott F, Ziem U, Reiffer FJ, Huth F, Ernst H, Mohr U [1987]. Carcinogenicity studies on fibres, metal compounds, and some other dusts in rats. Exp Pathol 2:129–152.

Potter RM [2000]. Method for determination of *in vitro* fiber dissolution rate by direct optical measurement of diameter decrease. Glastech Ber *73*:46–55.

Rice C, Heineman EF [2003]. An asbestos job exposure matrix to characterize fiber type, length, and relative exposure intensity. App Occup Environ Hyg *18*:506–512.

Riganti C, Aldieri E, Bergandi L, Tomatis M, Fenoglio I, Costamagna C, Fubini B, Bosia A, Ghigo D [2003]. Long and short fiber amosite asbestos alters at a different extent the redox metabolism in human lung epithelial cells. Toxicol Appl Pharmacol *193*:106–115.

Robinson BWS, Lake RA [2005]. Advances in malignant mesothelioma. N Engl J Med *353*: 1591–1603.

Robinson C, van Bruggen I, Segal A, Dunham M, Sherwood A, Koentgen F, Robinson BW, Lake RA [2006]. A novel SV40 TAg transgenic model of asbestos-induced mesothelioma: malignant transformation is dose dependent. Cancer Res *66*:10786–10794.

Rohs AM, Lockey JE, Dunning KK, Shukla R, Fan H, Hilbert T, Borton E, Wiot J, Meyer C, Shipley RT, Lemasters GK, Kapil V [2008]. Low-level fiber-induced radiographic changes caused by Libby vermiculite: a 25-year follow-up study. Am J Respir Crit Care Med *177*:630–637.

Rooker SJ, Vaughan NP, Le Guen JM [1982]. On the visibility of fibers by phase contrast microscopy. Am Ind Hyg Assoc J *43*:505–515.

Ross RM [2003]. The clinical diagnosis of asbestosis in this century requires more than a chest radiograph. Chest *124*:1120–1128.

Ross M, Virta RL [2001]. Occurrence, production and uses of asbestos. In: Nolan RP, Langer AM, Ross M, Wicks FJ, Martin RF, eds. The health effects of chrysotile asbestos: contribution of science to risk-management decisions. Can Mineral (Special Publication 5):79–88.

Ross M, Nolan RP, Nord GL [2007]. The search for asbestos within the Peter Mitchell Taconite iron ore mine, near Babbitt, Minnesota. Regul Toxicol Pharmacol *52*(Suppl 1):S43–S50.

Sakellariou K, Malamou-Mitsi V, Haritou A, Koumpaniou C, Stachouli C, Dimoliatis ID, Constantopoulos SH [1996]. Malignant pleural mesothelioma from nonoccupational asbestos exposure in Metsovo (north-west Greece): slow end of an epidemic? Eur Respir J *9*:1206–1210.

Satterley JD [2010]. Letter communication to the NIOSH Docket Office for Docket No. NIOSH–099C [http://www.cdc.gov/niosh/docket/archive/pdfs/NIOSH-099C/0099C-040910-Maimon_M_sub.pdf]. Date accessed: August 19, 2010.

Scherpereel A, Lee YC [2007]. Biomarkers for mesothelioma. Curr Opin Pulmon Med *13*: 339–443.

Schimmelpfeng J, Drosselmeyer E, Hofheinz V, Seidel A [1992]. Influence of surfactant components and exposure geometry on the effects of quartz and asbestos on alveolar macrophages. Environ Health Perspect *97*:225–231.

Schins RPF [2002]. Mechanisms of genotoxicity of particles and fibers. Inhal Toxicol *14*:57–78.

Schlesinger RB [1985]. Comparative deposition of inhaled aerosols in experimental animals and humans: a review. J Toxicol Environ Health *15*:197–214.

Scholze H, Conradt R [1987]. An *in vitro* study of the chemical durability of siliceous fibres. Ann Occup Hyg *31*:683–692.

Scott CC, Botelho RJ, Grinstein S [2003]. Phagosome maturation: a few bugs in the system. J Memb Biol 193:137–152.

Searl A [1994]. A review of the durability of inhaled fibres and options for the design of safer fibrils. Ann Occup Hyg 38:839–855.

Selevan SG, Dement JM, Wagoner JK, Froines JR [1979]. Mortality patterns among miners and millers of nonasbestiform talc: preliminary report. J Environ Pathol Toxicol 2:273–284.

Senyiğit A, Babayiğit C, Gökirmak M, Topçu F, Asan E, Coşkunsel M, Işik R, Ertem M [2000]. Incidence of malignant pleural mesothelioma due to environmental asbestos fiber exposure in the southeast of Turkey. Respiration 67:610–614.

Sesko A, Mossman B [1989]. Sensitivity of hamster tracheal epithelial cells to asbestiform minerals modulated by serum and by transforming growth factor β-1. Cancer Res 49:2743–2749.

Shatos MA, Doherty JM, Marsh JP, Mossman BT [1987]. Prevention of asbestos-induced cell death in rat lung fibroblasts and alveolar macrophages by scavengers of active oxygen species. Environ Res 44:103–116.

Shvedova AA, Kisin ER, Mercer R, Murray AR, Johnson VJ, Potapovich AI, Tyurina YY, Gorelik O, Arepalli S, Schwegler-Berry D, Hubbs AF, Antonini J, Evans DE, Ku B, Ramsey D, Maynard A, Kagan VE, Castranova V, Baron P [2005]. Unusual inflammatory and fibrogenic pulmonary responses to single-walled carbon nanotubes in mice. Am J Physiol Lung Cell Mol Physiol 289:L698–L708.

Shvedova AA, Kisin E, Murray AR, Johnson VJ, Gorelik O, Arepalli S, Hubbs AF, Mercer RR, Keohavong P, Sussman N, Jin J, Yin J, Stone S, Chen BT, Deye G, Maynard A, Castranova V, Baron PA, Kagan VE [2008]. Inhalation versus aspiration of single walled carbon nanotubes in C57BL/6 mice: inflammation, fibrosis, oxidative stress and mutagenesis. Am J Physiol Lung Cell Mol Physiol 295:L552–L565.

Siegel W, Smith AR, Greenburg L [1943]. The dust hazard in tremolite talc mining, including roentgenological findings in talc workers. Am J Roentogenol 49:11–29.

Siegrist HG, Wylie AG [1980]. Characterizing and discriminating the shape of asbestos particles. Environ Res 23:348–361.

Singh SV, Rahman Q [1987]. Interrelationship between hemolysis and lipid peroxidation of human erythrocytes induced by silicic acid and silicate dusts. J Appl Toxicol 7:91–96.

Smith WE, Hubert DD, Sobel HJ, Marquet E [1979]. Biologic tests of tremolite in hamsters. In: Dement JA, Lemen RA, eds. Dusts and disease. Park Forest South, IL: Pathtox Publishers, pp. 335–339.

Snipes MB [1996]. Current information on lung overload in nonrodent mammals: contrast with rats. In: Mauderly JL, McCunney RJ, eds. Particle overload in the rat lung and lung cancer: implications for human risk assessment. Proceedings of a conference held at the Massachusetts Institute of Technology, March 29–30, 1995. Washington, DC: Taylor and Francis, pp. 73–90.

Speit G [2002]. Appropriate *in vitro* test conditions for genotoxicity testing of fibers. Inhal Toxicol 14:79–90.

Spurny KR [1983]. Measurement and analysis of chemically changed mineral fibers after experiments *in vitro* and *in vivo*. Environ Health Perspect 51:343–355.

Stanton MF, Laynard M, Tegeris A, Miller E, May M, Kent E [1977]. Carcinogenicity of fibrous glass: pleural response in the rat in relation to fiber dimension. J Natl Cancer Inst 58:587–603.

Stanton MF, Layard M, Tegeris A, Miller E, May M, Morgan E, A Smith [1981]. Relation of particle dimension to carcinogenicity in amphibole asbestoses and other fibrous minerals. J Natl Cancer Inst 67:965–975.

Stayner LT, Dankovic D, Lemen RA [1996]. Occupational exposure to chrysotile asbestos and cancer risk: a review of the amphibole hypothesis. Am J Pub Health 86:179–186.

Stayner L, Kuempel E, Gilbert S, Hein M, Dement J [2007]. An epidemiologic study of the role of chrysotile asbestos fiber dimensions in determining respiratory disease risk in exposed workers. Occup Environ Med 65:613–619.

Stayner LT, Smith R, Bailer J, Gilbert S, Steenland K, Dement J, Brown D, Lemen R [1997]. Exposure-response analysis of respiratory disease risk associated with occupational exposure to chrysotile asbestos. Occup Environ Med 54:646–652.

Steenland K, Beaumont J, Halperin W [1984]. Methods of control for smoking in occupational cohort mortality studies. Scand J Work Environ Health 10:143–149.

Steenland K, Brown D [1995]. Mortality study of gold miners exposed to silica and nonasbestiform amphibole minerals: an update with 14 more years of followup. Am J Ind Med 27:217–229.

Stille WT, Tabershaw IR [1982]. The mortality experience of upstate New York talc workers. J Occup Med 24:480–484.

Straif K, Benbrahim-Tallaa L, Baan R, Grosse Y, Secretan B, El Ghissassi F, Bouvard V, Guha N, Freeman C, Galichet L, Cogliano V; WHO International Agency for Research on Cancer Monograph Working Group [2009]. A review of human carcinogens. Part C: metals, arsenic, dusts, and fibres. Lancet Oncol 10:453–454.

Su WC, Cheng Y [2005]. Deposition of fiber in the human nasal airway. Aerosol Sci Technol 29:888–901.

Sullivan P [2007]. Vermiculite, respiratory disease and asbestos exposure in Libby, Montana: update of a cohort mortality study. Environ Health Perspect 115:579–585.

Sussman RG, Cohen BS, Lippmann M [1991a]. Asbestos fiber deposition in human tracheobronchial cast. I: experimental. Inhal Toxicol 3:145–160.

Sussman RG, Cohen BS, Lippmann M [1991b]. Asbestos fiber deposition in human tracheobronchial cast. II: empirical model. Inhal Toxicol 3:161–178.

Suzuki Y, Yuen S [2001]. Asbestos tissue burden study on human malignant mesothelioma. Ind Health 39:150–160.

Suzuki Y, Yuen S [2002]. Asbestos fibers contributing to the induction of human malignant mesothelioma. Ann N Y Acad Sci 982:160–176.

Suzuki Y, Yuen S, Ashley R [2005]. Short thin asbestos fibers contribute to the development of human malignant mesothelioma: pathological evidence. Int J Hyg Environ Health 208:201–210.

Swain WA, O'Byrne KJ, Faux SP [2004]. Activation of p38 MAP kinase by asbestos in rat mesothelial cells is mediated by oxidative stress. Am J Physiol Lung Cell Mol Physiol 286:L859–L865.

Takeuchi T, Nakajima M, Morimoto K [1999]. A human cell system for detecting asbestos cytogenotoxicity *in vitro*. Mutat Res 438:63–70.

Taylor LE, Brown TJ, Benham AJ, Lusty PAJ, Minchin DJ [2006]. World mineral production 2000–2004. Keyworth, Nottingham, England: British Geological Survey.

Timbrell V [1982]. Deposition and retention of fibres in the human lung. Ann Occup Hyg 26:347–369.

Tossavainen A [2005]. World asbestos epidemic. Paper J1, Presented at The First International Occupational Hygiene Association (IOHA) International Scientific Conference (ISC) in Africa and IOHA 6th ISC, Pilanesberg, South Africa [http://www.saioh.org/ioha2005/Proceedings/Papers/SSJ/PaperJ1web.pdf]. Date accessed: November 20, 2006.

Tran CL, Buchanan D [2000]. Development of a biomathematical lung model to describe the exposure-dose relationship for inhaled dust among U.K. coal miners. Institute of Occupational Medicine Research Report TM/00/02. Edinburgh, UK: Institute of Occupational Medicine.

Unfried K, Schürkes C, Abel J [2002]. Distinct spectrum of mutations induced by crocidolite asbestos: clue for 8-hydroxydeoxyguanosine-dependent mutagenesis *in vivo*. Cancer Res 62:99–104.

U.S. Bureau of Mines [1996]. Dictionary of mining, mineral, and related terms. 2nd. ed. [http://xmlwords.infomine.com/USBM.htm]. Date accessed: December 18, 2009.

U.S. Senate [2007]. S. 742 Ban Asbestos in America Act of 2007 (Engrossed as Agreed to or Passed by Senate).

U.S. Geological Survey [2006]. Worldwide asbestos supply and consumption trends from 1900 through 2003 [http://pubs.usgs.gov/circ/2006/1298/c1298.pdf]. Date accessed: March 12, 2008.

U.S. Geological Survey [2007]. Mineral commodity summaries 2007; 199 pp [minerals.usgs.gov/minerals/pubs/mcs/2007/mcs2007.pdf]. Date accessed: March 12, 2008.

U.S. Geological Survey [2008]. Mineral commodity summaries 2008; 199 pp [minerals.usgs.gov/minerals/pubs/mcs/2008/mcs2008.pdf]. Date accessed: February 2, 2008.

Vallyathan V, Hanon N, Booth J, Schwegler D, Sepulveda M [1985]. Cytotoxicity of native and surface-modified asbestos. In: Beck EG, Bignon J, eds. *In vitro* effects of mineral dusts. Berlin-Heidelberg: Springer-Verlag. NATO ASI Series, Vol. G3, pp. 159–165.

Vallyathan V, Schwegler D, Reasor M, Stettler L, Clere J, Green FHY [1988]. Comparative *in vitro* cytotoxicity and relative pathogenicity of mineral dusts. Ann Occup Hyg 32:279–289.

Van Gosen BS [2007]. The geology of asbestos in the United States and its practical applications. Environ Eng Geosci 13:55–68.

Van Gosen BS [2006]. Reported historic asbestos mines, historic asbestos prospects, and natural asbestos occurrences in the Eastern United States. U.S. Geological Survey Open-File Report No. 2005–1189, Version 2.0; 1 p [http://pubs.usgs.gov/of/2005/1189/pdf/Plate.pdf]. Date accessed: August 18, 2010.

Van Gosen BS, Lowers HA, Sutley SJ, Gent CA [2004]. Using the geologic setting of talc deposits as an indicator of amphibole asbestos content. Env Geology 45:920–939.

Vastag E, Matthys H, Kohler D, Gronbeck G, Daikeler G [1985]. Mucociliary clearance and airways obstruction in smokers, ex-smokers and normal subjects who never smoked. Eur J Respir Dis 66:93–100.

Vianna NJ, Maslowsky J, Robert S, Spellman G, Patton B [1981]. Malignant mesothelioma: epidemiologic patterns in New York State. NY State J Med 81:735–738.

Vincent JH [2005]. Health-related aerosol measurement: a review of existing sampling criteria and proposals for new ones. Environ Monit 7:1037–1053.

Virta RL [2002]. Asbestos: U.S. Geological Survey Open-File Report No. 02-149, pp. 35 [http://pubs.usgs.gov/of/2002/of02-149/of02-149.pdf]. Date accessed: June 30, 2008.

Vu V, Barrett JC, Roycroft J, Schuman L, Dankovic D, Bbaro P, Martonen T, Pepelko W, Lai D [1996]. Chronic inhalation toxicity and carcinogenicity testing of respirable fibrous particles. Workshop report. Regul Toxicol Pharmacol 24:202–212.

Wagner CJ [1986]. Mesothelioma and mineral fibers: accomplishments in cancer research, 1985 prize year. Detroit, MI: General Motors Cancer Research Foundation, pp. 60–72.

Wagner JC, Skidmore JW, Hill RJ, Griffiths DM [1985]. Erionite exposure and mesotheliomas in rats. Br J Cancer 51:727–730.

Wagner GR, Attfield MD, Parker JE [1993]. Chest radiography in dust-exposed miners: promise and problems, potential and imperfections. Occup Med 8:127–141.

Wagner JC, Berry G, Timbrell V [1973]. Mesotheliomata in rats after inoculation with asbestos and other materials. Br J Cancer 28:173–175.

Wagner JC, Berry G, Skidmore JW, Timbrell V [1974]. The effects of the inhalation of asbestos in rats. Br J Cancer 29:252–269.

Wagner JC, Chamberlain M, Brown RC, Berry G, Pooley FD, Davies R, Griffiths DM [1982]. Biological effects of tremolite. Br J Cancer 45:352–360.

Walker AM, Loughlin JE, Friedlander ER, Rothman KJ, Dreyer NA [1983]. Projections of asbestos–related disease, 1980–2009. J Occup Med 25:409–425.

Wallace WE, Gupta NC, Hubbs AF, Mazza SM, Bishop HA, Keane MJ, Battelli LA, Ma J, Schleiff P [2002]. Cis-4-[F-18] fluoro-L-proline positron emission tomographic (PET) imaging of pulmonary fibrosis in a rabbit model. J Nucl Med 43:413–420.

Wallace WE, Keane MJ, Mike PS, Hill CA, Vallyathan V, Regad ED [1992]. Contrasting respirable quartz and kaolin retention of lecithin surfactant and expression of membranolytic activity following phospholipase A2 digestion. J Toxicol Environ Health 37:391–409.

Walton WH [1954]. Theory of size classification of airborne dust clouds by elutriation: the physics of particle size analysis. Br J Appl Phys 5(Suppl 3):s29–s37.

Walton WH [1982]. The nature, hazards and assessment of occupational exposure to airborne asbestos dust: a review. Part 1: nature and occurrence of asbestos dust. Ann Occup Hyg 25:121–154.

Wang Y, Faux SP, Hallden G, Kirn DH, Houghton CE, Lemoine NR, Patrick G [2004]. Interleukin-1β and TNF-α promote the transformation of human immortalised mesothelial cells by erionite. Int J Oncol 25:173–178.

Warheit D [1989]. Interspecies comparisons of lung responses to inhaled particles and gases. Crit Rev Toxicol 20:1–29.

Warheit DB, Overby LH, Gerwyn G, Brody AR [1988]. Pulmonary macrophages are attracted to inhaled particles through complement activation. Exp Lung Res *14*:51–66.

Watts JF, Wolstenholme J [2003]. An introduction to surface analysis by XPS and AES. New York: Wiley, pp. 224.

Webber J, Jackson KW, Parekh PP [2004]. Reconstruction of a century of airborne asbestos concentrations. Environ Sci Technol *38*:707–714.

Weill H [1994]. Biological effects: asbestos-cement manufacturing. Ann Occup Hyg *38*:533–538.

Weill H, Hughes JM, Churg AM [2004]. Changing trends in US mesothelioma incidence. Occup Environ Med *61*:438–441.

Weitzman SA, Graceffa P [1984]. Asbestos catalyzes hydroxyl and superoxide radical generation from hydrogen peroxide. Arch Biochem Biophys *228*:373–376.

Werner AJ, Hochella MF, Guthrie GD, Hardy JA, Aust AE, Rimstidt JD [1995]. Asbestiform riebeckite (crocidolite) dissolution in the presence of Fe chelators: implications for mineral-induced disease. Am Mineral *80*:1093–1103.

WHO [1997]. Determination of airborne fibre number concentrations; a recommended method, by phase contrast optical microscopy (membrane filter method). Geneva: World Health Organization.

Woodworth C, Mossman B, Craighead J [1983]. Induction of squamous metaplasia in organ cultures of hamster trachea by naturally occurring and synthetic fibers. Cancer Res *43*:4906–4912.

Wylie AG [1988]. The relationship between the growth habit of asbestos and the dimensions of asbestos fibers. Presented at the SME Annual Meeting, Phoenix, AZ, January 25–28.

Wylie AG, Bailey KF [1992]. The mineralogy and size of airborne chrysotile and rock fragments: ramifications of using the NIOSH 7400 method. Am Ind Hyg Assoc J *53*:442–447.

Wylie AG, Virta RL, Russek E [1985]. Characterizing and discriminating airborne amphibole cleavage fragments and amosite fibers: implications for the NIOSH method. Am Ind Hyg Assoc J *46*:197–201.

Wylie AG, Virta RL, Segreti JM [1987]. Characterization of mineral population by index particle: implications for the Stanton hypothesis. Environ Res *43*:427–439.

Wylie AG, Bailey KF, Kelse JW, Lee RJ [1993]. The importance of width in asbestos fiber carcinogenicity and its implications for public policy. Am Ind Hyg Assn J *54*:239–252.

Wylie AG, Skinner HC, Marsh J, Snyder H, Garzione C, Hodkinson D, Winters R, Mossman BT [1997]. Mineralogical features associated with cytotoxic and proliferative effects of fibrous talc and asbestos on rodent tracheal epithelial and pleural mesothelial cells. Toxicol Appl Pharmacol *147*:143–150.

Yamaguchi R, Hirano T, Ootsuyama Y, Asami S, Tsurudome Y, Fukada S, Yamato H, Tsuda T, Tanaka I, Kasai H [1999]. Increased 8-hydroxyguanine in DNA and its repair activity in hamster and rat lung after intratracheal instillation of crocidolite asbestos. Jpn J Cancer Res *90*:505–509.

Yang H, Bocchetta M, Kroczynska B, Elmishad AG, Chen Y, Liu Z, Bubici C, Mossman BT, Pass HI, Testa JR, Franzoso G, Carbone M [2006]. TNF-α inhibits asbestos-induced cytotoxicity via a NF-κB-dependent pathway, a possible mechanism for asbestos-induced oncogenesis. Proc Natl Acad Sci USA *103*:10397–10402.

Yano E, Wang Z, Wang X [2001]. Cancer mortality among workers exposed to amphibole-free chrysotile asbestos. Am J Epidemiol *154*:538–543.

Yu CP, Asgharian B, Yen BM [1986]. Impaction and sedimentation deposition. Am Ind Hyg Assoc J *47*:72–77.

Zeidler-Erdely PC, Calhoun WJ, Amercedes BT, Clark MP, Deye GJ, Baron P, Jones W, Blake T, Castranova V [2006]. *In vitro* cytotoxicity of Manville Code 100 glass fibers: effect of fiber length on human alveolar macrophages. Part Fibre Toxicol *3*:5–11.

Zhou Y, Su WC, Cheng YS [2007]. Fiber deposition in the tracheobronchial region: experimental measurements. Inhal Toxicol *19*:13, 1071–1078.

Zodac P [1940]. A talc quarry near Chester, Vermont. Rocks and Minerals *15*:369–371.

Zoltai T [1981]. Amphibole asbestos mineralogy. In: Veblen DR, ed. Amphiboles and other hydrous pyriboles. Rev Mineral *9A*:237–278.

Zumwalde RD, Ludwig HR, Dement JM [1981]. Industrial hygiene report, Homestake Mining Company, Lead, South Dakota. Cincinnati, OH: U.S. Department of Health and Human Services, Centers for Disease Control, National Institute for Occupational Safety and Health. NTIS #PB85-243640, pp. 255.

6 Glossary

6.1 NIOSH Definition of Potential Occupational Carcinogen

Potential Occupational Carcinogen means any substance, or combination or mixture of substances, which causes an increased incidence of benign and/or malignant neoplasms, or a substantial decrease in the latency period between exposure and onset of neoplasms in humans or in one or more experimental mammalian species as the result of any oral, respiratory or dermal exposure, or any other exposure which results in the induction of tumors at a site other than the site of administration. This definition also includes any substance which is metabolized into one or more potential occupational carcinogens by mammals.

6.2 Definitions of New Terms Used in this Roadmap

Countable elongate mineral particle: A particle that meets specified dimensional criteria and is to be counted according to an established protocol. A countable elongate mineral particle under the NIOSH REL for *Airborne Asbestos Fibers and Related Elongate Mineral Particles* is any asbestiform fiber, acicular or prismatic crystal, or cleavage fragment of a *covered mineral* that is longer than 5 μm and has a minimum aspect ratio of 3:1 on the basis of a microscopic analysis of an air sample using NIOSH Method 7400 or an equivalent method.

Covered mineral: A mineral encompassed by a specified regulation or recommended standard. Under the NIOSH REL for Airborne Asbestos Fibers and Related Elongate Mineral Particles, covered minerals include those minerals having the crystal structure and elemental composition of the asbestos varieties (chrysotile, riebeckite asbestos [crocidolite], cummingtonite-grunerite asbestos [amosite], anthophyllite asbestos, tremolite asbestos, and actinolite asbestos) or their nonasbestiform analogs (the serpentine minerals antigorite and lizardite and the amphibole minerals contained in the cummingtonite-grunerite mineral series, the tremolite-ferroactinolite mineral series, and the glaucophane-riebeckite mineral series).

Elongate mineral particle (EMP): Any mineral particle with a minimum aspect ratio of 3:1. This Roadmap is focused on EMPs that are of inhalable, thoracic, or respirable size, as described below in Section 6.2.

Elongate particle (EP): Any particle with a minimum aspect ratio of 3:1. The research described in this Roadmap is focused on EPs that are of inhalable, thoracic, or respirable size, as described below in Section 6.2.

6.3 Definitions of Inhalational Terms

Inhalable particulate matter: Particles that deposit anywhere in the respiratory tract.

This varies by species, but for humans such matter can be approximated as those particles captured according to the following collection efficiency, regardless of sampler orientation with respect to wind direction:

$$IPM(d_{ae}) = 0.5\,(1 + \exp[-0.06\,d_{ae}]) \pm 10;\text{ for } 0 < d_{ae} \leq 100\ \mu m$$

Where $IPM(d_{ae})$ = the collection efficiency and d_{ae} is the aerodynamic diameter in μm. [ACGIH 1999]

Respirable particulate matter: Particles that deposit anywhere in the gas-exchange region of the lung. This varies by species, but for humans such matter can be approximated as those particles captured according to the following collection efficiency: $RPM(d_{ae}) = IPM(d_{ae})[1-F(x)]$

Where $F(x)$ = cumulative probability function of the standardized normal variable, x.

$x = \ln(d_{ae}/4.25\ \mu m)/\ln(1.5)$ [ACGIH 1999]

Thoracic particulate matter: Particles that deposit anywhere within the lung airways and the gas-exchange region. This varies by species, but for humans such matter can be approximated as those particles captured according to the following capture efficiency: $TPM(d_{ae}) = IPM(d_{ae})[1-F(x)]$

Where $F(x)$ = cumulative probability function of the standardized normal variable, x

$x = \ln(d_{ae}/11.64\ \mu m)/\ln(1.5)$ [ACGIH 1999]

6.4 Definitions of General Mineralogical Terms and Specific Minerals

Definitions from several sources are provided in the following tables for many of the mineralogical terms used in this Roadmap. Definitions of some terms in this table differ. Also, definitions of these same terms, as used by various authors whose work has been cited in this Roadmap, may vary from those provided here. It is not possible to know and/or provide each of the variant definitions.

6.5 References for Definitions of General Mineralogical Terms, Specific Minerals, and Inhalational Terms

ACGIH (American Conference of Governmental Industrial Hygienists) [1999]. Particle size–selective sampling for particulate air contaminants. Vincent JH, ed. Cincinnati, OH: ACGIH.

American Geological Institute [2005]. Glossary of Geology, 5th ed. Neuendorf KKE, Mehl JP, Jackson JA, eds. Alexandria, VA: American Geological Institute.

Leake BE, Woolley AR, Arps CES, Birch WD, Gilbert CM, Grice JD, Hawthorne FC, Kato A, Kisch HF, Krivovichev VG, Linthout K, Laird J, Mandarino JA, Maresch WV, Nickel EH, Rock NMS, Schumacher JC, Smith DC, Stephenson NCN, Ungaretti L, Whittaker EJW, Youzhi G [1997]. Nomenclature of the amphiboles: report of the Subcommittee on Amphiboles of the International Mineralogical Association Commission on new minerals and mineral names. Can Mineral 35:219–246.

NIOSH [1990a]. Comments of the National Institute for Occupational Safety and Health on the Occupational Safety and Health Administration's Notice of Proposed Rulemaking on Occupational Exposure to Asbestos, Tremolite, Anthophyllite, and Actinolite. OSHA Docket

No. H-033d, April 9, 1990 [http://www.cdc.gov/niosh/review/public/099/pdfs/Asbestos-Testimony_April%209_1990.pdf]. Date accessed: December 17, 2009.

U.S. Bureau of Mines [1996]. Dictionary of mining, mineral, and related terms. 2nd ed [http://xmlwords.infomine.com/USBM.htm]. Date accessed: December 18, 2009.

Table 1. Definitions of general mineralogical terms

Term	Dictionary of Mining, Mineral, and Related Terms [U.S. Bureau of Mines 1996] [Note: Footnotes identify the Primary Source Citation for the definition]	Glossary of Geology, 5th ed. [American Geological Institute 2005]	Leake et al. [1997]	NIOSH [1990a]
Acicular[1]	1. A mineral consisting of fine needlelike crystals, e.g., natrolite. 2. Slender needlelike crystal. 3. Refers to needlelike crystals.[2]	[crystal]: Said of a crystal that is needlelike in form.		
Amphibole	A mineral group; characterized by double chains of silica tetrahedra having the composition $A_{0-1}B_2Y_5Z_8O_{22}(OH,F,Cl)$, where (A = Ca,Na,K,Pb,B), (B = Ca,Fe,Li,Mg,Mn,Na), (Y = Al,Cr,Fe,Mg,Mn,Ti), and (Z = Al,Be,Si,Ti); in the orthorhombic or monoclinic crystal systems, including actinolite, anthophyllite, arfvedsonite, cummingtonite, hornblende, richterite, glaucophane, grunerite, anthophyllite, riebeckite, tremolite, and others. All display a diagnostic prismatic cleavage in two directions parallel to crystal faces and intersecting at angles of about 54° and 124°. Some members may be asbestiform.	1. A group of dark [sic] rock-forming ferromagnesian silicate minerals, closely related in crystal form and composition and having the general formula: $A_{2-3}B_5(Si,Al)_8O_{22}(OH)_2$, where A = Mg, Fe^{2+}, Ca, or Na, and B = Mg, Fe^{2+}, Fe^{3+}, Li, Mn, or Al. It is characterized by a cross-linked double chain of tetrahedral with silicon:oxygen ratio of 4:11, by columnar or fibrous prismatic crystals, and by good prismatic cleavage in two directions parallel to the crystal faces and intersecting at angles of 56° and 124°; colors range from white to black. Most amphiboles crystallize in the monoclinic system, some in the orthorhombic. They constitute an abundant and widely distributed constituent in igneous and metamorphic rocks (some are wholly metamorphic), and they are analogous in chemical composition to the pyroxenes. 2. A mineral of the amphibole group, such as hornblende, anthophyllite, cummingtonite, tremolite, actinolite, riebeckite, glaucophane, arfvedsonite, etc.	A mineral comprising a double silicate chain with the general formula $AB_2{}^{VI}C_5{}^{IV}T_8O_{22}(OH)_2$ with the components of the formula conventionally described as A, B, C, T and "OH" corresponding to the following crystallographic sites: A one site per formula unit; B two M4 sites per formula unit; C a composite of five sites made up of 2 M1, 2 M2 and 1 M3 sites per formula unit; T eight sites, in two sets of four, that need not be distinguished; "OH" two sites per formula unit. The ions considered normally to occupy these sites are in the following categories: (empty site) and K at A only; Na at A or B; Ca at B only; L-type ions: Mg, Fe^{2+}, Mn^{2+}, Li and rarer ions of similar size, at C or B; M-type ions: Al at C or T, Fe^{3+} and, more rarely Mn^{3+}, Cr^{3+} at C only; high-valency ions: Ti^{4+} at C or T, Zr^{4+} at C only, Si at T only; anions: OH, F, Cl, O at "OH". M-type ions normally occupy M2 sites and so are normally limited to two of the five C sites.	Minerals in the amphibole group are widely distributed in the earth's crust in many igneous or metamorphic rocks. In some instances, the mineral deposits contain sufficient quantities of the asbestiform minerals to be economically minable for commercial use. The minerals and mineral series of the amphibole group have variable compositions with extensive elemental substitutions. They are found in forms ranging from massive to fibrous. The most common commercially exploited asbestiform varieties of this mineralogical group include crocidolite, amosite, anthophyllite, tremolite, and actinolite. Crocidolite, amosite, and anthophyllite are selectively mined for commercial use, whereas tremolite and actinolite are most often found as a contaminant in other mined commodities such as talc and vermiculite. The amphiboles have good thermal and electrical insulation properties, and they have moderate to good resistance to acids.

See footnotes at end of table.

(Continued)

Table 1. Definitions of general mineralogical terms (Continued)

Term	Dictionary of Mining, Mineral, and Related Terms [U.S. Bureau of Mines 1996] [Note: Footnotes identify the Primary Source Citation for the definition]	Glossary of Geology, 5th ed. [American Geological Institute 2005]	Leake et al. [1997]	NIOSH [1990a]
Amphibole (Continued)		3. A term sometimes used a synonym for hornblende. Etymol: Greek "amphibolos", "ambiguous, doubtful", in reference to its many varieties.	Exceptions may occur to the above "normal" behavior. Four groups are classified depending on the occupancy of the B sites: Mg-Fe-Mn-Li group; calcic group; sodic-calcic group; and sodic group. Asbestiform amphiboles should be named according to their precise mineral name (when known) followed by the suffix—asbestos, e.g., anthophyllite-asbestos, tremolite-asbestos.	
Asbestiform[1]	1. Said of a mineral that is fibrous, i.e., like asbestos.	Said of a mineral that is composed of separable fibers.		A specific type of mineral fibrosity in which the growth is primarily in one dimension and the crystals form naturally as long, flexible fibers. Fibers can be found in bundles that can be easily separated into smaller bundles or ultimately into fibrils.

See footnotes at end of table.

(Continued)

Table 1. Definitions of general mineralogical terms (Continued)

Term	Dictionary of Mining, Mineral, and Related Terms [U.S. Bureau of Mines 1996] [Note: Footnotes identify the Primary Source Citation for the definition]	Glossary of Geology, 5th ed. [American Geological Institute 2005]	Leake et al. [1997]	NIOSH [1990a]
Asbestos[1]	1. A commercial term applied to silicate minerals that separate readily into thin, strong fibers that are flexible, heat resistant, and chemically inert, thus making them suitable for uses (as in yarn, cloth, paper, paint, brake linings, tiles, insulation, cement, fillers, and filters) where incombustible, nonconducting, or chemically resistant material is required. Since the early 1970's, there have been serious environmental concerns about the potential health hazards of asbestos products, which has resulted in strong environmental regulations. 2. Any asbestiform mineral of the serpentine group (chrysotile, best adapted for spinning and the principal variety in commerce) or amphibole group (esp. actinolite, anthophyllite, gedrite, cummingtonite, grunerite, riebeckite, and tremolite). 3. A term strictly applied to asbestiform actinolite.	1. A commercial term applied to a group of silicate minerals that readily separate into thin, strong fibers that are flexible, heat resistant, and chemically inert, and are therefore suitable for uses (as in yarn, cloth, paper, paint, brake linings, tiles, insulation, cement, fillers, and filters) where incombustible, nonconducting, or chemically resistant material is required. 2. A mineral of the asbestos group [sic], principally chrysotile (best adapted for spinning) and certain fibrous varieties of amphibole (esp. amosite, anthophyllite, and crocidolite). 3. A term strictly applied to the fibrous variety of actinolite. Certain varieties are deleterious to health.		Asbestos is a generic term for a number of silicate minerals with a fibrous crystalline structure. The quality of commercially used asbestos depends on the mineralogy of the asbestiform variety, the degree of fiber development, the ratio of fibers to acicular crystals or other impurities, and the length and flexibility of the fibers. The asbestiform varieties of these minerals can be found in both the amphibole and serpentine mineral groups. The asbestiform varieties occur in veins or small veinlets within rock containing or composed of the common (nonasbestiform) variety of the same mineral. The major asbestiform varieties of minerals used commercially are chrysotile, tremolite-actinolite asbestos, cummingtonite-grunerite asbestos, anthophyllite asbestos, and crocidolite. Asbestos is marketed by its mineral name (e.g., anthophyllite asbestos), its variety name (e.g., chrysotile or crocidolite), or its trade name (e.g., Amosite).

(Continued)

See footnotes at end of table.

Table 1. Definitions of general mineralogical terms (Continued)

Term	Dictionary of Mining, Mineral, and Related Terms [U.S. Bureau of Mines 1996] [Note: Footnotes identify the Primary Source Citation for the definition]	Glossary of Geology, 5th ed. [American Geological Institute 2005]	Leake et al. [1997]	NIOSH [1990a]
Cleavage fragment[1]		A fragment of a crystal that is bounded by cleavage faces.		A fragment produced by the breaking of crystals in directions that are related to the crystal structure and are always parallel to possible crystal faces. Minerals with perfect cleavage can produce perfect regular fragments. Amphiboles with prismatic cleavage will produce prismatic fragments. *Note:* These fragments can be elongated and may meet the definition of a fiber upon microscopic examination.
Crystal habit	The forms typically appearing on specimens of a mineral species or group, rarely all the forms permitted by its point group. Crystal habits range from highly diverse, e.g., calcite, to almost never showing crystal faces, e.g. turquoise. In addition to describing mineral habits with form names, e.g. prismatic, pyramidal, or tetrahedral, other names for appearances are used, e.g. fibrous, columnar, platy, or botryoidal. Intergrowths are given by specific description.[3]	The general shape of crystals, e.g. cubic, prismatic, fibrous. For a given type of crystal, the habit may vary from locality to locality depending on environment of growth.		

See footnotes at end of table.

(Continued)

Table 1. Definitions of general mineralogical terms (Continued)

Term	Dictionary of Mining, Mineral, and Related Terms [U.S. Bureau of Mines 1996] [Note: Footnotes identify the Primary Source Citation for the definition]	Glossary of Geology, 5th ed. [American Geological Institute 2005]	Leake et al. [1997]	NIOSH [1990a]
Fiber[1]	The smallest single strand of asbestos or other fibrous material.[4]	A strengthening cell, usually elongated, tapering, and thick-walled, occurring in various parts of vascular plants. [Note: The definition provided does not refer to mineral fibers.]		An acicular single crystal or similarly elongated polycrystalline aggregate particles. Such particles have macroscopic properties such as flexibility, high aspect ratio, silky luster, and axial lineation. These particles have attained their shape primarily because of manifold dislocation planes that are randomly oriented in two axes but parallel in the third. *Note*: Upon microscopic examination, only particles that have a 3:1 or greater aspect ratio are defined as fibers. Other macroscopic properties used to define fibers cannot be ascertained for individual particles examined microscopically.
Fibril[1]	1. A single fiber which cannot be separated into smaller components without losing its fibrous properties or appearance.[5]			A single fiber that cannot be separated into smaller components without losing its fibrous properties or appearances.
Fibrous[1]	1. Applied to minerals that occur as fibers, such as asbestos. Syn: asbestiform 2. Consisting of fine threadlike strands, e.g., satin spar variety of gypsum.			

See footnotes at end of table.

(Continued)

Table 1. Definitions of general mineralogical terms (Continued)

Term	Dictionary of Mining, Mineral, and Related Terms [U.S. Bureau of Mines 1996] [Note: Footnotes identify the Primary Source Citation for the definition]	Glossary of Geology, 5th ed. [American Geological Institute 2005]	Leake et al. [1997]	NIOSH [1990a]
Fibrous habit		The tendency of certain minerals, e.g., asbestos, to crystallize in needlelike grains or fibers.		
Fibrous structure	If the crystals in a mineral aggregate are greatly elongated and have a relatively small cross-section, the structure or texture is fibrous. The fibers may be parallel, as in crocidolite and sometimes in calcite and cerussite. When the fibers are very fine, they may impart a silky luster to the aggregate, as in crocidolite or satin-spar gypsum. There is also a feltlike type. Fibrous crystals may radiate from a center, forming asteriated or starlike groups, either coarse or fine, as frequently observed in pyrolusite, wavellite, natrolite and tourmaline, and sometimes in stibnite and other minerals. Also called fibrous texture.[6]	Fibrous prismatic structure: A prismatic structure in which each first-order prism is like a simple prism in showing nonspherulitic prismatic and noncomposite prismatic substructure, but the prisms have much higher length/width ratios than typical simple prisms, occurring as long fibers.		
Fibrous texture	In mineral deposits, a pattern of finely acicular, rod-like crystals, e.g., in chrysotile and amphibole asbestos.[7]		In mineral deposits, a pattern of finely acicular, rod-like crystals, e.g., in chrysotile and amphibole asbestos.	

See footnotes at end of table.

(Continued)

Table 1. Definitions of general mineralogical terms (Continued)

Term	Dictionary of Mining, Mineral, and Related Terms [U.S. Bureau of Mines 1996] [Note: Footnotes identify the Primary Source Citation for the definition]	Glossary of Geology, 5th ed. [American Geological Institute 2005]	Leake et al. [1997]	NIOSH [1990a]
Mineral	1. A naturally occurring inorganic element or compound having an orderly internal structure and characteristic chemical composition, crystal form, and physical properties. CF: metallic. 2. In miner's phraseology, ore. See also: ore. 3. See: mineral species; mineral series; mineral group. 4. Any natural resource extracted from the earth for human use; e.g, ores, salts, coal, or petroleum. 5. In flotation, valuable mineral constituents of ore as opposed to gangue minerals. 6. Any inorganic plant or animal nutrient. 7. Any member of the mineral kingdom as opposed to the animal and plant kingdoms.[7]	1. A naturally occurring inorganic element or compound having a periodically repeating arrangement of atoms and characteristic chemical composition, resulting in distinctive physical properties. 2. An element or chemical compound that is crystalline and formed as a result of geologic processes. Materials formed by geological processes from artificial substances are no longer accepted (after 1995) as new minerals (Nickel, 1995). Mercury, a liquid, is a traditional exception to the crystallinity rule. Water is not a mineral (although ice is), and crystalline biological and artificial materials are not minerals (cf. mineraloid). 3. Any naturally formed inorganic material, i,e., a member of the mineral kingdom as opposed to the plant or animal kingdom.		A homogeneous, naturally occurring, inorganic crystalline substance. Minerals have distinct crystal structures and variation in chemical composition, and are given individual names.

See footnotes at end of table.

(Continued)

Table 1. Definitions of general mineralogical terms (Continued)

Term	Dictionary of Mining, Mineral, and Related Terms [U.S. Bureau of Mines 1996] [Note: Footnotes identify the Primary Source Citation for the definition]	Glossary of Geology, 5th ed. [American Geological Institute 2005]	Leake et al. [1997]	NIOSH [1990a]
Mineral series				A mineral series includes two or more members of a mineral group in which the cations in secondary structural position are similar in chemical properties and can be present in variable but frequently limited ratios (e.g., cummingtonite-actinolite). The current trend in referring to a mineral series is to simplify long series names by using the mineral name of only one (end or intermediate) member (i.e., tremolite-actinolite-ferroactinolite).
Mineral variety				The variety distinguishes minerals that are conspicuously different from (1) those considered normal within the common crystallization I habits, polytypes, and other structural variants, and (2) those with different physical properties such as color. Varieties are named by mineralogists, miners, gemologists, manufacturers of industrial products, and mineral collectors.
Needle	5. A needle-shaped or acicular mineral crystal.	[crystal]: A needle-shaped or acicular mineral crystal.		

See footnotes at end of table.

(Continued)

Table 1. Definitions of general mineralogical terms (Continued)

Term	Dictionary of Mining, Mineral, and Related Terms [U.S. Bureau of Mines 1996] [Note: Footnotes identify the Primary Source Citation for the definition]	Glossary of Geology, 5th ed. [American Geological Institute 2005]	Leake et al. [1997]	NIOSH [1990a]
Nonasbestiform habit				Each of the six commercially exploited asbestiform minerals also occurs in a nonasbestiform mineral habit. These minerals have the same chemical formula as the asbestiform variety, but they have crystal habits where growth proceeds in two or three dimensions instead of one dimension. When milled, these minerals do not break into fibrils but rather into fragments resulting from cleavage along the two or three growth planes. Particles thus formed are referred to as cleavage fragments and can meet the definition of a fiber for regulatory purposes.
Prism	1. An open crystal form with faces and their intersecting edges parallel to the principle crystallographic axis. Prisms have three (trigonal), four (tetragonal), six (ditrigonal or hexagonal), eight (ditetragonal), or twelve (dihexagonal) faces. The nine-sided prisms of tourmaline are a combination of trigonal and hexagonal prisms.	[crystal] A crystal form having three, four, six, eight, or twelve faces, with parallel intersection edges, and which is open only at the two ends of the axis parallel to the intersection edges of the faces.		

See footnotes at end of table.

(Continued)

Table 1. Definitions of general mineralogical terms (Continued)

Term	Dictionary of Mining, Mineral, and Related Terms [U.S. Bureau of Mines 1996] [Note: Footnotes identify the Primary Source Citation for the definition]	Glossary of Geology, 5th ed. [American Geological Institute 2005]	Leake et al. [1997]	NIOSH [1990a]
Prismatic	2. Pertaining to a crystallographic prism. 3. Descriptive of a crystal with one dimension markedly longer than the other two. 4. Descriptive of two directions of cleavage.	[crystal] Said of a crystal that shows one dimension markedly longer than the other two.		
Serpentine Minerals		A rock consisting almost wholly of serpentine-group minerals, e.g., antigorite, chrysotile, or lizardite, derived from the hydration of ferromagnesian silicate minerals such as olivine and pyroxene. Accessory chlorite, talc, and magnetite may be present.		The serpentine minerals belong to the phyllosilicate group of minerals. The commercially important variety is chrysotile, which originates in the asbestiform habit. Antigorite and lizardite are two other types of serpentine minerals that are structurally distinct. The fibrous form of antigorite is called picrolite.

See footnotes at end of table.

(Continued)

Table 1. Definitions of general mineralogical terms (Continued)

Term	Dictionary of Mining, Mineral, and Related Terms [U.S. Bureau of Mines 1996] [Note: Footnotes identify the Primary Source Citation for the definition]	Glossary of Geology, 5th ed. [American Geological Institute 2005]	Leake et al. [1997]	NIOSH [1990a]
Zeolite	1. A generic term for class of hydrated silicates of aluminum and either sodium or calcium or both, of the type $Na_2O \cdot Al_2O_3 \cdot nSiO_2 \cdot xH_2O$. The term originally described a group of naturally occurring minerals. The natural zeolites are analcite, chabazite, heulandite, natrolite, stilbite, and thomsonite. Artificial zeolites are made in a variety of forms, ranging from gelatinous to porous and sandlike, and are used as gas adsorbents and drying agents as well as water softeners. Both natural and artificial zeolites are used extensively for water softening. The term zeolite now includes such diverse groups of compounds as sulfonated organics or basic resins, which act in a similar manner to effect either cation or anion exchange.	a generic term for a large group of white or colorless (sometimes tinted red or yellow by impurities) hydrous aluminosilicate minerals that have an open framework structure of interconnected $(Si,Al)O_4$ tetrahedra with exchangeable cations and H_2O molecules in structural cavities. They have a ratio of $(Al + Si)$ to nonhydrous oxygen of 1:2, and are characterized by their easy and reversible loss of water of hydration and by their ready fusion and swelling when strongly heated under the blowpipe. Zeolites have long been known to occur as well formed crystals in cavities in basalt. Of more significance is their occurrence as authigenic minerals in the sediments of saline lades and the deep sea and esp. in beds of tuff. They form "during and after burial, generally by reaction of pore waters with solid aluminosilicate materials (e.g., volcanic glass, feldspar, biogenic silica, and clay minerals)."[8]		

See footnotes at end of table.

(Continued)

Table 1. Definitions of general mineralogical terms (Continued)

Term	Dictionary of Mining, Mineral, and Related Terms [U.S. Bureau of Mines 1996] [Note: Footnotes identify the Primary Source Citation for the definition]	Glossary of Geology, 5th ed. [American Geological Institute 2005]	Leake et al. [1997]	NIOSH [1990a]
Zeolite (Continued)	2. A group of hydrous aluminosilicates that are similar to the feldspars. They easily lose and regain their water of hydration and they fuse and swell when heated. Zeolites are frequently used in water softening, ion exchange and absorbent applications.			

[1]Additional definitions can be found in Lowers H, Meeker G [2002]. Tabulation of asbestos-related terminology. Open-file report 02-458 [http://pubs.usgs.gov/of/2002/ofr-02-458/], pp. 70. Date accessed: December 21, 2009.
[2]Nelson A [1965]. Dictionary of mining. New York: Philosophical Library, pp. 523.
[3]Pryor EJ [1963]. Dictionary of mineral technology. London: Mining Publications, Ltd., pp. 437.
[4]Mersereau SF [1947]. Materials of industry, 4th ed. New York: McGraw-Hill, pp. 623.
[5]Campbell, WJ et al. Selected silicate minerals and their asbestiform varieties. USBM Circular 8751. Washington, DC: U.S. Department of the Interior, Bureau of Mines.
[6]Chamber's mineralogical dictionary [1948]. New York: Chemical Publishing Co., pp. 47.
[7]American Geological Institute [1987]. Glossary of geology, 3rd ed. Alexandria, VA; pp. 788. In addition, also by the American Geological Institute: Glossary of geology and related sciences [1957], pp. 325; supplement [1969], pp. 72.
[8]Hay RL [1978]. Geologic occurrence of zeolites. In: Sand LB, Mumpton FA, eds. Natural zeolites. New York: Pergamon, pp. 135–143.

Table 2. Definitions of specific minerals

Term	Dictionary of Mining, Mineral, and Related Terms [U.S. Bureau of Mines 1996] [Note: Footnotes identify the Primary Source Citation for the definition]	Glossary of Geology, 5th ed. [American Geological Institute 2005]	Leake et al. [1997]	NIOSH [1990a]
Actinolite	A monoclinic mineral, $2[Ca_2(Fe,Mg)_5Si_8O_{22}(OH)_2]$; in the hornblende series $Mg/(Mg+Fe^{2+}) = 0.50$ to 0.89 of the amphibole group; forms a series with tremolite; green, bladed, acicular, fibrous (byssolite asbestos), or massive (nephrite jade); prismatic cleavage; in low-grade metamorphic rocks.	A bright-green or grayish-green monoclinic mineral of the amphibole group: $Ca_2(Fe,Mg)_5(OH)_2[Si_8O_{22}]$. It may contain manganese. It sometimes occurs in the form of asbestos, and also in fibrous, radiated, or columnar forms in metamorphic rocks (such as schists) and in altered igneous rocks.	A monoclinic calcic amphibole intermediate between ferroactinolite and tremolite: $Ca_2(Fe,Mg)_5Si_8O_{22}(OH)_2$; with $Mg/(Mg+Fe^{2+})$ between 0.5 and 0.9 (otherwise if $\leq$ 0.5 it is ferroactinolite, and if $\geq$ 0.9 it is tremolite).	Actinolite can occur in both the asbestiform and nonasbestiform mineral habits and is in the mineral series tremolite-ferroactinolite.[1] The asbestiform variety is often referred to as actinolite asbestos.
Amosite	1. A monoclinic mineral in the cummingtonite-grunerite series.[2] 2. A commercial asbestos composed of asbestiform gedrite, grunerite, or anthophyllite of the amphibole group; has typically long fibers.	A commercial term for an iron-rich, asbestiform variety of amphibole occurring in long fibers. It may consist of an orthorhombic amphibole (anthophyllite or gedrite) or of a monoclinic amphibole (cummingtonite or grunerite).		Amosite is the commercial term derived from the acronym "Asbestos Mines of South Africa." Amosite is in the mineral series cummingtonite-grunerite,[1] in which both asbestiform and nonasbestiform habits of the mineral can occur. This mineral type is commonly referred to as "brown asbestos."
Antigorite	A monoclinic mineral, $(Mg,Fe)_3Si_2O_5(OH)_4$; kaolinite-serpentine group; polymorphous with clinochrysotile, lizardite, orthochrysotile, parachrysotile; greasy variegated green; used as an ornamental stone.	A macroscopically lamellar brown to green monoclinic serpentine mineral, which consists structurally of alternating wave forms in which the 1:1 T-O layer reverses sides and direction of curvature at each wave null point. In most specimens the repeat distance of the wave pattern measures between 25.5 and 51.0 Å: $(Mg, Fe^{2+})_3Si_2O_3(OH)_4$.		

See footnotes at end of table.

(Continued)

Table 2. Definitions of specific minerals (Continued)

Term	Dictionary of Mining, Mineral, and Related Terms [U.S. Bureau of Mines 1996] [Note: Footnotes identify the Primary Source Citation for the definition]	Glossary of Geology, 5th ed. [American Geological Institute 2005]	Leake et al. [1997]	NIOSH [1990a]
Anthophyllite	An orthorhombic mineral, $4[Mg,Fe)_7Si_8O_{22}(OH)_2]$; amphibole group; commonly lamellar or fibrous, green to clove-brown; in schists from metamorphosed ultramafic rocks; a nonspinning grade of asbestos.	A clove-brown to colorless orthorhombic mineral of the amphibole group: $(Mg,Fe^{2+})_2(Mg,Fe^{2+})_5Si_8O_{22}(OH)_2$. It is dimorphous with cummingtonite; with increase in aluminum it grades into gedrite. Anthophyllite occurs in metamorphosed ultrabasic rocks, typically with olivine or talc or in monomineralic aggregates of parallel or radiating asbestiform fibers. It has been mined for asbestos.	An orthorhombic Mg-Fe-Mn-Li amphibole: $Mg_7Si_8O_{22}(OH)_2$; may also contain divalent iron but with $Mg/(Mg+Fe^{2+}) \geq 0.50$ (otherwise ferro-anthophyllite), and with Si > 7.00 (otherwise it is gedrite).	Anthophyllite can occur in both the asbestiform and nonasbestiform mineral habits. The asbestiform variety is often referred to as anthophyllite asbestos.
Attapulgite	A light-green, magnesium-rich clay mineral, named from its occurrence at Attapulgus, GA, where it is quarried as fuller's earth. Crystallizes in the monoclinic system.	palygorskite		
Byssolite	An olive-green asbestiform variety of tremolite-actinolite.	An olive-green asbestiform variety of tremolite-actinolite.		
Clinoptilolite	A monoclinic mineral, $(Na,K,Ca)_2Al_3(Al,Si)_2Si_{13}O_{36} \cdot 12H_2O$; of the zeolite group.	A group name for a monoclinic zeolite mineral with the general formula: $A_{2-3}(Si,Al)_{18}O_{36} \cdot 11H_2O$, where A = Na, K, or Ca.		
Chrysotile	A monoclinic mineral (clinochrysotile), or orthorhombic mineral (orthochrysotile, parachrysotile), $[Mg_6(OH)_8Si_4O_{10}]$; serpentine group; forms soft, silky white, yellow, green, or gray flexible fibers as veins in altered ultramafic rocks; the chief asbestos minerals. (Not to be confused with chrysolite.)	A white, gray, or greenish orthorhombic or monoclinic mineral of the serpentine group: $Mg_3(OH)_4Si_2O_5$. It is a highly fibrous, silky variety of serpentine, and constitutes the most important type of asbestos. Not to be confused chrysolite.		Chrysotile generally occurs segregated as parallel fibers in veins or veinlets and can easily separate into individual fibers or bundles. Often referred to as "white asbestos," it is used commercially for its good spinnability in the making of textile products, and as an additive in cement' or friction products.

See footnotes at end of table.

(Continued)

Table 2. Definitions of specific minerals (Continued)

Term	Dictionary of Mining, Mineral, and Related Terms [U.S. Bureau of Mines 1996] [Note: Footnotes identify the Primary Source Citation for the definition]	Glossary of Geology, 5th ed. [American Geological Institute 2005]	Leake et al. [1997]	NIOSH [1990a]
Crocidolite	An asbestiform variety of riebeckite; forms lavender-blue, or indigo-blue, or leek-green silky fibers and massive and earthy forms; suited for spinning and weaving. Also spelled krokitolit.	An asbestiform variety of riebeckite; forms lavender-blue, or indigo-blue, or leek-green silky fibers and massive and earthy forms. Also spelled krokidolit.		Crocidolite is from the fibrous habit of the mineral riebeckite and is in the mineral series glaucophane-riebeckite, in which both asbestiform and nonasbestiform habits can occur. This mineral type is commonly referred to as "blue asbestos."
Cummingtonite	A monoclinic mineral, $(Fe,Mg)_7Si_8O_{22}(OH)_2$; amphibole group; has $Mg/(Mg+Fe^{2+}) = 0.30$ to 0.69; prismatic cleavage; may be asbestiform; in amphibolites and dacites; fibrous varieties (amosite, magnesium rich, and montasite, iron rich) are used as asbestos.	A dark green, brown, gray, or beige monoclinic member of the amphibole group: $(Mg,Fe^{2+})_7Si_8O_{22}(OH)_2$. It is dimorphous with anthophyllite, and typically contains calcium and manganese. Cummingtonite occurs in metamorphosed ironstone, mafic and ultrabasic rocks, some dacites and rhyolites, and as a component of uralite. Its iron-rich variety is grunerite.	A monoclinic Mg-Fe-Mn-Li amphibole: $Mg_7Si_8O_{22}(OH)_2$; may also contain divalent iron but with $Mg/(Mg+Fe^{2+}) \geq 0.50$ (otherwise it is grunerite)	
Erionite		A white hexagonal zeolite mineral. [Ed. Note: Designated as Erionite (Ca,K,Na) depending on the dominant cation substitution.		
Ferroactinolite	A monoclinic mineral, $Ca_3(Fe^{2+},Mg)_5Si_8O_{22}(OH)_2]$; amphibole group; has $Mg/(Mg+Fe^{2+}) = 0$ to 0.50; forms a series with tremolite and actinolite. Formerly called ferrotremolite.	A green-black monoclinic mineral component representing a theoretical end-member of the amphibole group: $Ca_2Fe^{2+}_5Si_8O_{22}(OH)_2$. Syn: ferrotremolite.	A monoclinic calcic amphibole: $Ca_2Fe^{2+}_5Si_8O_{22}(OH)_2$; may also contain magnesium but with $Mg/(Mg+Fe^{2+}) \leq 0.5$ (otherwise it is actinolite).	

See footnotes at end of table.

(Continued)

Table 2. Definitions of specific minerals (Continued)

Term	Dictionary of Mining, Mineral, and Related Terms [U.S. Bureau of Mines 1996] [Note: Footnotes identify the Primary Source Citation for the definition]	Glossary of Geology, 5th ed. [American Geological Institute 2005]	Leake et al. [1997]	NIOSH [1990a]
Fluoro-edenite		A vitreous dark brown monoclinic mineral of the amphibole group: (Na,K)Ca$_2$(Mg,Fe^{2+})$_5$(Si$_7$Al)O$_{22}$(F,OH). It represents edenite with F>OH.		
Grunerite	A monoclinic mineral, (Fe,Mg)$_7$Si$_8$O$_{22}$(OH)$_2$; amphibole group; with Mg/(Mg+Fe^{2+}) = 0-0.30; forms series with cummingtonite and magnesiocummingtonite; fibrous or needlelike, commonly in radial aggregates; characteristic of iron formations in the Lake Superior and Labrador Trough regions. Also spelled gruenerite.		A monoclinic Mg-Fe-Mn-Li amphibole: Fe^{2+}$_7$Si$_8$O$_{22}$(OH)$_2$; may also contain magnesium but with Mg/(Mg+Fe^{2+}) < 0.50 (otherwise it is cummingtonite).	
Halloysite	1. A monoclinic mineral, 2[Al$_4$Si$_4$(OH)$_8$O$_{10}$]; kaolinite-serpentine group; made up of slender tubes as shown by electron microscopy; a gangue mineral in veins. 2. Used as a group name to include natural "halloysite minerals" with different levels of hydration, as well as those formed artificially.	A 1:1 aluminosilicate clay mineral Al$_2$Si$_2$O$_5$(OH)$_4$·X(H$_2$O) similar to kaolinite but perhaps with some Al(IV) and interlayer cations to compensate for the Al(IV). Probably because of this it is able to incorporate water in the interlayer space [Bailey 1989]. The terms "halloysite (7Å)" and halloysite (10Å)" were recommended for the anhydrous and dihydrate forms, respectively [Brindley and Pegro 1976][3]; the term "endellite" should not be used [Bailey et al. 1980][4].		

See footnotes at end of table.

(Continued)

Table 2. Definitions of specific minerals (Continued)

Term	Dictionary of Mining, Mineral, and Related Terms [U.S. Bureau of Mines 1996] [Note: Footnotes identify the Primary Source Citation for the definition]	Glossary of Geology, 5th ed. [American Geological Institute 2005]	Leake et al. [1997]	NIOSH [1990a]
Lizardite	A trigonal and hexagonal mineral, $Mg_3Si_2O_5(OH)_4$; kaolinite-serpentine group; polymorphous with antigorite, clinochrysotile, orthochrysotile, and parachrysotile; forms a series with nepouite; in platy masses as an alteration product of ultramafic rocks; the most abundant serpentine mineral.	The most abundant form of the trioctahedral serpentine minerals. It crystallizes as flat platelets. Variable amounts of Al substitute for both Mg and Si in the ideal serpentine formula of $Mg_3Si_2O_5(OH)_4$ to create a better lateral fit between the component octahedral and tetrahedral sheets than found in antigorite and chrysotile. Several polytypes exist: rhombohedral, trigonal, hexagonal, or monoclinic.		
Mordenite	A white, yellowish, or pinkish member of the zeolite group of minerals with the formula $(Ca,Na_2,K_2)Al_2Si_{10}O_{24} \cdot 7H_2O$.	A white, yellowish, or pinkish orthorhombic zeolite mineral: $(Na_2,Ca,K_2)Al_2Si_{10}O_{24} \cdot 7H_2O$.		
Palygorskite	1. A monoclinic and orthorhombic mineral, $(OH)_2(Mg,Al)_4(Si,Al)_8O_{20} \cdot 8H_2O$; fibrous; in desert soils. 2. A general name for lightweight fibrous clay minerals showing significant substitution of aluminum for magnesium; characterized by distinctive rodlike shapes under an electron microscope.	(a) A white, grayish, yellowish, or grayish-green chain-structure clay mineral: $(Mg,Al)_2Si_4O_{10}(OH) \cdot 4H_2O$. It crystallizes in several monoclinic and orthorhombic polytypes. (b) A group name for monoclinic minerals with an analogous composition, but with Mg replaced by Mn or Na, and AL replaced by Fe^{3+} or Mn^{3+}.		

See footnotes at end of table.

(Continued)

Table 2. Definitions of specific minerals (Continued)

Term	Dictionary of Mining, Mineral, and Related Terms [U.S. Bureau of Mines 1996] [Note: Footnotes identify the Primary Source Citation for the definition]	Glossary of Geology, 5th ed. [American Geological Institute 2005]	Leake et al. [1997]	NIOSH [1990a]
Phillipsite	A monoclinic mineral, $(K,Na,Ca)_{1-2}(Si,Al)_8O_{16} \cdot 6H_2O$; zeolite group; commonly occurs in complex twinned crystals; in basalt amydules, in pelagic red clays, in palagonite tuffs, in alkaline saline lakes from silicic vitric volcanic ash, in alkaline soils, and around hot springs in Roman baths.	A colorless or white monoclinic zeolite mineral. Usually designated as phillipsite – (Ca, K, or Na) depending on which is the dominant exchangeable cation, $(Ca,K,Na)_2(Si,Al)_8O_{16} \cdot 6H_2O$.		
Richterite		A brown, yellow, or rose-red monoclinic member of the amphibole group: $Na_2CaMg_5Si_8O_{22}(OH)_2$. Cf: soda tremolite	A monoclinic sodic-calcic amphibole: $Na(CaNa)Mg_5Si_8O_{22}(OH)_2$; may also contain divalent iron but with $Mg/(Mg+Fe^{2+}) \geq 0.5$ (otherwise it is ferrorichterite)	
Riebeckite	A monoclinic mineral, $Na_2Ca(Mg,Fe^{2+})_5Si_8O_{22}(OH)_2$ [sic]; amphibole group with $Mg/(Mg+Fe^{2+}) = 0$ to 0.49 and $Fe^{3+}/(Fe^{3+}+Al) = 0.7$ to 1.0; forms a series with magnesioriebeckite; fibrous; in soda-rich rhyolites, granites, and pegmatites; crocidolite variety is blue asbestos; tiger eye is crocidolite replaced by quartz.	A dark blue or black monoclinic mineral of the amphibole group: $Na_2Fe^{2+}{}_3Fe^{3+}{}_2Si_8O_{22}(OH)_2$. It occurs as a primary constituent in some acid or sodium-rich igneous rocks. See also: crocidolite	A monoclinic sodic amphibole: $Na_2(Fe^{2+},Fe^{3+}{}_2)Si_8O_{22}(OH)_2$; may also contain aluminum in place of trivalent iron but with $^{VI}Al < Fe^{3+}$ otherwise it is ferroglaucophane, and may also contain sodium and potassium in the A position but with $(Na+K)_A < 0.50$ otherwise it is arfvedsonite, and may also contain magnesium in place of divalent iron but with $Mg/(Mg+Fe^{2+}) < 0.5$ otherwise it is magnesioriebeckite	

See footnotes at end of table.

(Continued)

Table 2. Definitions of specific minerals (Continued)

Term	Dictionary of Mining, Mineral, and Related Terms [U.S. Bureau of Mines 1996] [Note: Footnotes identify the Primary Source Citation for the definition]	Glossary of Geology, 5th ed. [American Geological Institute 2005]	Leake et al. [1997]	NIOSH [1990a]
Sepiolite	A monoclinic mineral, $Mg_2Si_6O_{15}(OH)_2 \cdot 6H_2O$; soft; sp gr, 2, but fibrous dry masses float on water; occurs in veins in calcite and in alluvial deposits formed from weathering of serpentine masses, chiefly in Asia Minor, as meerschaum; may be used in making pipes, ornamental carvings.	An orthorhombic chain-structure clay mineral: $Mg_4Si_6O_{15}(OH)_2 \cdot 6H_2O$. It is a white to light gray or light yellow material, extremely lightweight, absorbent, and compact, that is found chiefly in Asia Minor and is used for making tobacco pipes, cigar and cigarette holders and ornamental carvings. Sepiolite occurs in veins with calcite, and in alluvial deposits formed from weathering of serpentine masses.		
Tremolite	A monoclinic mineral, $2[Ca_2Mg_5Si_8O_{22}(OH)_2]$; amphibole group with magnesium replaced by iron, and silicon by aluminum toward actinolite; white to green, long-bladed or stout prismatic crystals, may show columnar, fibrous, or granular masses or compact aggregates; in low-grade metamorphic rocks such as dolomitic limestones and talc schists; the nephrite variety is the gemstone jade; the asbestiform variety is byssolite.	A white to dark-gray monoclinic mineral of the amphibole group: $Ca_2Mg_5Si_8O_{22}(OH)_2$. It has varying amounts of iron, and may contain manganese and chromium. Tremolite occurs in long blade-shaped or short stout prismatic crystals and also in columnar, fibrous, or granular masses or compact aggregates, generally in metamorphic rocks such as crystalline dolomitic limestones and talc schists. It is a constituent of much commercial talc.	A monoclinic calcic amphibole: $Ca_2Mg_5Si_8O_{22}(OH)_2$; may also contain divalent iron but with $Mg/(Mg+Fe^{2+}) \geq 0.9$ (otherwise it is actinolite)	Tremolite can occur in both the asbestiform and nonasbestiform mineral habits and is in the mineral series tremolite-ferroactinolite.[1] The asbestiform variety is often referred to as tremolite asbestos.
Winchite		A blue or gray monoclinic member of the amphibole group: $NaCa(Mg,Al)Si_8O_{22}(OH)_2$	A monoclinic sodic-calcic amphibole: $(CaNa)Mg_4(Al,Fe^{3+})Si_8O_{22}(OH)_2$; may also contain divalent iron but with $Mg/(Mg+Fe^{2+}) \geq 0.5$ (otherwise it is ferrorichterite)	

See footnotes at end of table.

(Continued)

Table 2. Definitions of specific minerals (Continued)

Term	Dictionary of Mining, Mineral, and Related Terms [U.S. Bureau of Mines 1996] [Note: Footnotes identify the Primary Source Citation for the definition]	Glossary of Geology, 5th ed. [American Geological Institute 2005]	Leake et al. [1997]	NIOSH [1990a]
Wollastonite	A triclinic mineral of the pyroxenoid group: CaSiO$_3$. It is dimorphous with parawollastonite. Wollastonite is found in contact-metamorphosed limestones, and occurs usually in cleavable masses or sometimes in tabular twinned crystals; it may be white, gray, brown, red, or yellow. It is not a pyroxene. Symbol, Wo.	A triclinic or monoclinic chain silicate mineral of the pyroxenoid type: CaSiO$_3$. It [*in the original source, a word is missing here.*] dimorphous with parawollastonite. Wollastonite is found in contact-metamorphosed limestones, and occurs usually in cleavable masses or sometimes in tabular twinned crystals; it may be white, gray, brown, red, or yellow. It is not a pyroxene. Several polytypes have been characterized. Symbol: Wo.		

[1]Mineral series such as cummingtonite-grunerite and tremolite-ferroactinolite are created when one cation is replaced by another in a crystal structure without significantly altering the structure. There may be a gradation in the structure in some series, and minor changes in physical characteristics may occur with elemental substitution. Usually a series has two end members with an intermediate substitutional compound being separately named, or just qualified by being referred to as members of the series. Members of the tremolite-ferroactinolite series are hydroxylated calcium-magnesium, magnesium-iron, and iron silicates, with the intermediate member of this series being named actinolite.
[2]Sinclair WE [1959]. Asbestos: its origin, production and utilization. Mining, 2nd ed. London: Publications, Ltd. pp. 512.
[3]Brindley GW, Pedro G [1976]. Meeting of the nomenclature committee of AIPEA; Mexico City, July 12, 1975. AIPEA Newsletter 12:5–6.
[4]The original source does not provide the full reference for this citation.

www.ingramcontent.com/pod-product-compliance
Lightning Source LLC
Chambersburg PA
CBHW080247180526
45167CB00006B/2448